How Things Work

Around and around, wheels called gears revolve as their edges mesh. One gear turns another . . . and another . . . and another. These gears keep a printing press rolling. Other gear systems provide power for a variety of machines, including ten-speed bicycles (pages 6–11) and wind-up alarm clocks (pages 12–13).

MICHAEL MELFORD/THE IMAGE BANK

BOOKS FOR WORLD EXPLORERS
NATIONAL GEOGRAPHIC SOCIETY

Contents

JEFFREY R. WERNER/AFTER-IMAGE, INC.

DAVID P. JOHNSON, N.G.S. STAFF

DAVID FORBERT/SHOSTAL ASSOCIATES, INC.

ANDY LEVIN/BLACK STAR

MICHAEL KELLER/ALPHA

COVER: *A neon sign spells out* HOW THINGS WORK. *Gases called neon and argon make the letters glow. In the clear glass tubes, neon shows its natural color—orange-red. Argon helps color the frosted tube green. You can read more about neon signs and how they work on pages 32 through 35.*

N.G.S. PHOTOGRAPHER JOSEPH LAVENBURG

N.G.S. PHOTOGRAPHER JOSEPH H. BAILEY

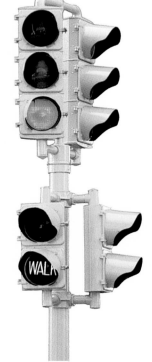

JAMES A. SUGAR AND PHILIP R. LEONHARDI, N.G.S. STAFF

JON T. SCHNEEBERGER, N.G.S. STAFF

Copyright © 1983 National Geographic Society
Library of Congress CIP data: p. 104

Introduction

If you want to have fun, you might take a bike ride or play a video game. If you have to study, you may reach for a pocket calculator. Every day, you use objects that make your life easier and happier. But do you know what goes on inside these things? In this book, you will explore the inner workings of many familiar devices—and a few you may never have seen.

Before you begin this journey of discovery, it would help you to know an important fact: All things work according to natural laws. Scientists group these laws into a field they call physics. What is physics? It is the science that studies matter—or substances—and energy. In other words, physics examines what things are made of and how they perform work. The laws of physics show how matter and energy are related. By learning to use these laws, people have greatly changed their lives. Without an understanding of physics, they could not build automobiles. There would be no radios. Electric lights and telephones would not exist.

Forces at work

Two principles of physics will help you understand how different things work. First, forces change the motion and the shape of an object. You are perhaps most familiar with the force of gravity. Gravity tries to pull soaring sailplanes, hot-air balloons, and the space shuttle back to earth. Other forces include such things as magnetism and electricity.

Second, different kinds of energy can produce these forces. You already are familiar with many sources of energy. For instance, energy from your leg muscles turns a bicycle wheel. Energy from a wound-up spring runs a clock. Energy from burning fuel powers a rocket.

Scientists divide the study of physics into several areas. These areas include electricity and magnetism; heat; sound, or acoustics (uh-KOO-stix); mechanics;

light, or optics (OP-tix); and the structure of atoms—the tiny particles that form the basic building blocks of all things. This book will touch upon all these areas. With some subjects, such as the bicycle, you will learn about just one part, the gears. With other subjects, such as the vending machine, you will look into several different areas, such as the coin acceptor section and the product delivery section. With still other objects, such as the camera, you will learn about almost every system.

Electricity and magnetism

Many devices rely on electric currents and electric and magnetic fields. Electricity and magnetism are related. In fact, electric currents create magnetic fields. Knowing this, scientists now can produce powerful magnets by sending an electric current through a wire coiled around a piece of iron or steel. Such magnets are called electromagnets. Scientists think that electric currents

deep within the earth make the earth itself behave like an enormous magnet. Electric currents and magnetic fields play important roles in making many objects work. Compasses and electric motors are only a couple of things that depend on magnets and electromagnets.

Electromagnetism is important in other ways, too. Radio waves, TV waves, microwaves, radar, visible light, and X rays all are forms of electromagnetic energy. They travel at an extremely high rate of speed and can even move through a vacuum, or empty space. These waves flow much like the waves in the ocean do. Each wave has a high point, or crest, and a low point, or trough. The distance between crests is called the wavelength. The number of crests and troughs that pass a given point each second is called the frequency of that wave. The various kinds of electromagnetic waves differ in wavelength and in frequency, but in a vacuum they move at the same speed.

Force and motion

The study of the motion of objects and the forces that act on them is called mechanics. *How Things Work* explains the mechanical energy that makes an alarm clock ring at the time you want it to and that enables bicycle gears to help you go farther while using less force.

Heat power

Heat is another important area of physics. Scientists call the study of heat thermodynamics (ther-mo-dye-NAM-iks). Heat is one of the most useful forms of energy. Heated air rises above cooler air. The energy created by rising air helps hot-air balloons operate. Heat causes certain substances to expand. In a medical thermometer, for example, heat in your body causes a liquid metal called mercury to expand inside a glass tube. A scale on the tube measures how far the mercury expands and tells you, in degrees, how much heat energy your body contains at that time. Heat also produces electric energy in steam-powered generators. It produces mechanical energy in internal combustion engines—the engines that power most automobiles.

Energy in sound

Sound waves enable you to enjoy the notes created by musical instruments. The study of sound, or acoustics, explains what sound is and how it moves. Sound is vibration. It travels in waves, but not in the same way that electromagnetic waves travel. Unlike electromagnetic waves, which can move through empty space, sound waves must pass through some kind of material. This material can be a solid, a liquid, or a gas. You can hear only those sound waves that lie within a certain range of frequencies, from about 20 to 20,000 vibrations a second. Waves of higher or lower frequencies cannot be heard as sound. But they can damage your ears if they are of high intensity.

Putting light to work

The field of optics, or light, includes mainly those electromagnetic waves that you can see. Visible light helps many devices work. Smoke detectors, for example, sound an alarm when smoke reflects a light beam onto a light-sensitive cell inside them. The blockage causes an electric current to flow. Lasers create super-powerful light beams that can cut through metals or can carry billions of telephone signals at one time.

How things work

As you will see, *How Things Work* includes both simple machines and more complicated ones. Read about them, but don't experiment yourself without help from an adult. If, at first, you have trouble understanding some of the principles explained in these chapters, read them over again. The more you learn about the forces that make these devices work, the better you'll be able to understand the world around you.

How a Ten-speed Bicycle Works

Every time you pedal down the street on a ten-speed bicycle, you control one of the most popular means of travel in the United States today. Since you supply the power for the bicycle's wheels, gears, levers, and cables, you are the bike's engine. The gears help you make the best use of your power. They change muscle energy into mechanical energy.

The earliest bicycles had no gears or steering systems—they didn't even have pedals. The rider sat on a seat between two wheels and "walked" the bicycle. In 1861, designers added pedals to the front wheel. Thousands of people bought these newfangled bikes.

The sport of bicycle racing soon became popular. A bicycle known as the "ordinary" gave the racers speed. The rider perched atop a huge front wheel. Each time the rider moved the pedals around once, the wheel made a complete turn. The larger the wheel was, the farther the bike went. So manufacturers began making bikes with front wheels as tall as five feet (1½ m).* But the ordinary was dangerous, because riders often fell off.

People wanted a safer bicycle. Soon designers found that, by adding a chain and a rear sprocket—a chain-catching wheel—they could reduce the size of the front wheel without affecting the bicycle's speed. This new bike was called the "safety."

By the turn of the century, some safeties had variable-speed gears. The gears were contained inside the hub, or center bar, of the rear wheel. A rider could control the bike's speed by changing gears.

Bikes of today look similar to the safety, but they have more gears. On the next pages, you'll see how the gears on a ten-speed bike work—and how you can put them to work for you.

*Metric figures in this book have been rounded off.

Cyclists pedal through the countryside during a week-long ride held every summer in Iowa. Thousands make the trip. A newspaper, the Des Moines Register, organizes the event.

BILL GILLETTE/AFTER-IMAGE, INC.

A New Yorker pedals in a parade on a bicycle called the "ordinary" (left). The ordinary was the most popular form of transportation on the road a hundred years ago. In the late 1800s, few cars existed. Bicycles provided the most efficient and inexpensive way of traveling. The ordinary's front wheel could be as large as 5 feet ($1^1/_2$ m) in diameter. Each time the pedals were pushed around, the big wheel made a complete turn. This enabled the bicycle to cover more distance faster than smaller bikes. However, the rider teetered high off the ground and often pitched off when the wheel hit a bump. Later, the "safety" bicycle replaced the ordinary. It had two wheels of about the same size. Bicycles of today are patterned after the safety.

Fans watch as a racer zips by on a multispeed bike (below). The rider chooses from many different gears to suit his strength, the wind direction, and the kind of ground he's riding over. High gears give him the greatest speed. Most racers pump the pedals as many as a hundred times a minute. At that rate, their bikes can travel as fast as 30 miles an hour (48 km/h). Racers always wear safety helmets. Even the most skilled cyclist can take a tumble.

How gears change

When you ride a ten-speed bike, the pedals turn two chainwheels—metal wheels ringed with gear teeth. A chain links one or the other of these chainwheels to one of five sprockets on the rear wheel. The rear sprockets turn the rear wheel. The chain, one chainwheel, and one rear sprocket make up one gear. When you change gears, you move the chain to a sprocket that will change the number of times the rear wheel goes around each time you turn the pedals once. In the highest gear, the rear wheel may turn three times for each turn of the pedals. In the lowest gear, the rear wheel makes only $1^{1}/_{2}$ turns, but pedaling is easier.

In the large drawing at right, you can see the full power train. The chain (green) connects the chainwheels to the group of rear sprockets called the freewheel. On this bike, the chain links the small chainwheel (tan) and the largest freewheel sprocket (orange). In this position the bike is in the lowest gear. You would use this gear for climbing very steep hills. The larger the freewheel sprocket, the fewer times the rear wheel turns each time you push the pedals around—and the easier it is to pedal.

Freewheel

Cable to rear shift lever

REAR DERAILLEUR

Jockey pulley

Tension pulley

A device called the rear derailleur (dih-RAY-lur) moves the chain from one rear sprocket to another (left). To move the chain onto the smallest sprocket (purple), the rider pushes a lever near the handlebars (see page 10) all the way forward. This creates slack on the cable leading to the rear derailleur. A spring inside the derailleur (not shown) pushes the derailleur away from the wheel. The derailleur pulls the chain from a larger sprocket to the small one. Two pulleys help the chain move. The top pulley, called the jockey pulley, helps to guide the chain. The tension pulley, below, moves forward or back to take up the slack. Ghost images show the derailleur's position for the highest gear.

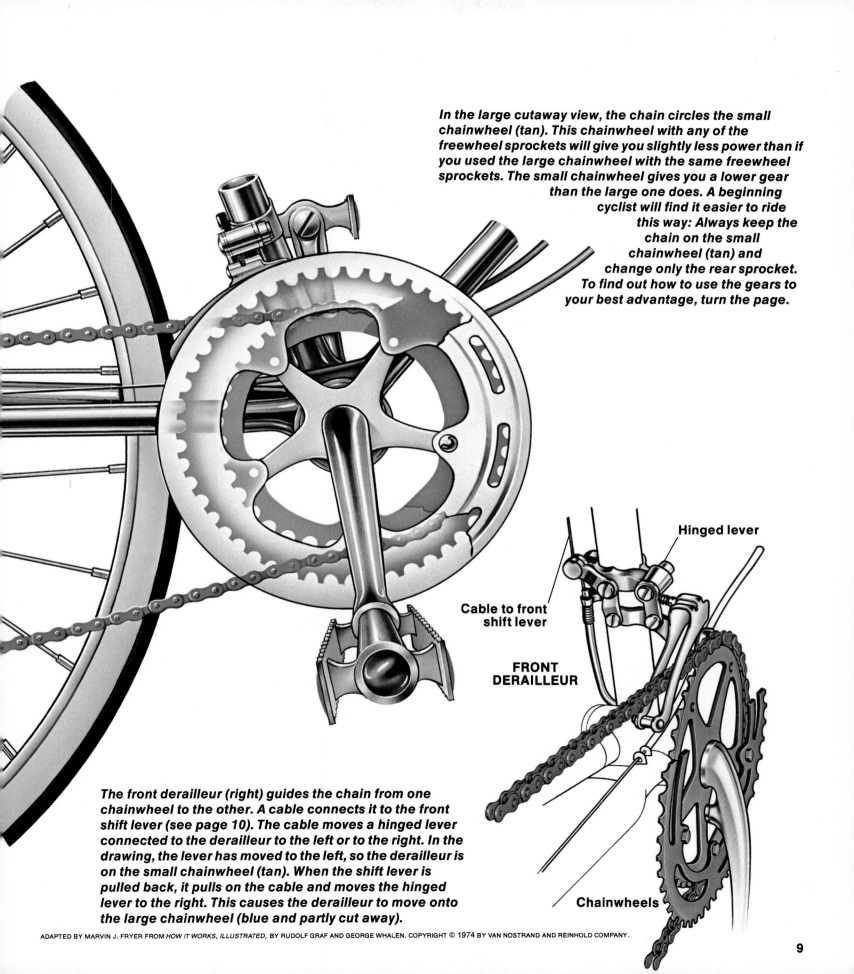

In the large cutaway view, the chain circles the small chainwheel (tan). This chainwheel with any of the freewheel sprockets will give you slightly less power than if you used the large chainwheel with the same freewheel sprockets. The small chainwheel gives you a lower gear than the large one does. A beginning cyclist will find it easier to ride this way: Always keep the chain on the small chainwheel (tan) and change only the rear sprocket. To find out how to use the gears to your best advantage, turn the page.

Hinged lever

Cable to front shift lever

FRONT DERAILLEUR

The front derailleur (right) guides the chain from one chainwheel to the other. A cable connects it to the front shift lever (see page 10). The cable moves a hinged lever connected to the derailleur to the left or to the right. In the drawing, the lever has moved to the left, so the derailleur is on the small chainwheel (tan). When the shift lever is pulled back, it pulls on the cable and moves the hinged lever to the right. This causes the derailleur to move onto the large chainwheel (blue and partly cut away).

Chainwheels

ADAPTED BY MARVIN J. FRYER FROM HOW IT WORKS, ILLUSTRATED, BY RUDOLF GRAF AND GEORGE WHALEN. COPYRIGHT © 1974 BY VAN NOSTRAND AND REINHOLD COMPANY.

Presto, chango! With a flick of a lever, a ten-speed bicycle's gear system goes into action. This drawing (right) shows the shift levers as the cyclist would see them. The lever on the left controls the front derailleur. The lever on the right controls the rear derailleur. When you push the left lever forward, the chain moves onto the small chainwheel (tan circle). When you pull the left lever back, the chain moves onto the large chainwheel (blue circle). As you move the right lever from front to back, the chain is gradually moved from the smallest rear sprocket (purple circle) to the largest (orange circle). The drawing next to the levers shows a cyclist's-eye view of the chain in all ten gears. Gears make pedaling easy for Andy Fonda, 13, as he rides his ten-speed up a hill (below). His friend David Patrick, 9, must stand on the pedals of his single-speed bike to get extra push. Both boys live in Manassas, Virginia.

MARVIN J. FRYER (RIGHT)

WILLIAM R. FONDA (BELOW)

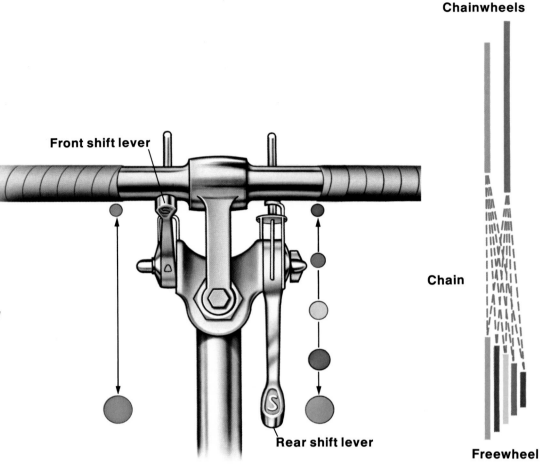

Front shift lever

Rear shift lever

Chainwheels

Chain

Freewheel

Using your gears

Would you like to know how much pedaling power each gear actually gives you? This chart (left) gives the gear ratios of a typical ten-speed bike with a 27-inch (69-cm) rear wheel. That is, it tells how far the rear wheel moves with each turn of the pedals. To find a gear ratio, first divide the number of teeth on a chainwheel by the number of teeth on a rear sprocket. Then multiply the answer by the size of the rear wheel. To use the chart, look at the number of teeth on each rear sprocket. They are listed in the column at the far left. The other two columns list the number of teeth on the small chainwheel (42) and the number on the large chainwheel (52). When the chain is on the largest rear sprocket (28 teeth) and the small chainwheel (42 teeth) the gear ratio is about 40 (42 ÷ 28 x 27). That means the rear wheel will move about 40 inches (102 cm) during one turn of the pedals. The drawing of the bicycle next to that line indicates that the gear should be used on the steepest inclines. Higher gear ratios can produce greater speed, but they make the bike harder to pedal.

Number of teeth on small chainwheel	Number of teeth on large chainwheel
42	52

Number of teeth on rear sprockets	Number of inches traveled by the rear wheel with each turn of the pedals	
28	40	50
24	47	58
20	56	70
17	66	82
14	81	100

2

How Alarm Clocks Work

You may have an electric clock-radio that wakes you to music, or you may use a simpler wind-up alarm clock like the one shown here. The power for wind-up alarms comes from a metal coil called a mainspring (opposite). You wind the mainspring by turning a key (A). As the spring unwinds, it supplies power to move the hands.

How does it do this? Inside, the clock has a series of large and small wheels with toothed edges. The mainspring (B) is attached to a large wheel called the main wheel (C). As the mainspring unwinds, the main wheel turns with it. The teeth of the main wheel mesh with the teeth of the next wheel in the series. All the wheels fit together in this way. As the main wheel turns, each wheel turns the wheel next to it.

Wheels of different sizes take different amounts of time to make a full revolution. The wheels are arranged so that the hour hand and the minute hand turn at different speeds to keep the right time. The whole series of moving wheels is called the wheel train (green).

Without something to stop it, the mainspring would unwind as soon as you finished winding it. A mechanism called the escapement (orange) acts as a brake. It keeps the power of the mainspring from "escaping." The escapement consists of three parts: a balance wheel (D), a tiny metal spring called a hairspring (E), and a lever (F). The hairspring controls the motion of the balance wheel. When the balance wheel turns one way, the hairspring tightens. Then the hairspring unwinds, pushing the balance wheel back the other way. That makes the balance wheel rock to and fro. As the balance wheel rocks, it pushes the lever back and forth. When the lever moves, two pins on the end of the lever take turns catching the teeth of the orange wheel. This wheel is connected to the last wheel of the wheel train (G) and slows down the whole wheel train by making it stop and go. You hear this action as *tick-tock tick-tock*.

In this combination of two photographs (left), you can see both the outside of a wind-up alarm clock and the works inside. This kind of clock is powered by a mainspring. To wind up the spring, you turn a key (1) on the back of the clock (above). To set the alarm, you pull out the alarm-set knob (2) and turn it. This lets you line up the alarm hand on the front of the clock with the time you want the alarm to go off. Then you push the knob back in. Next, you pull out the alarm control button (3). Now the alarm will go off at the time you set.

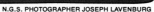

N.G.S. PHOTOGRAPHER JOSEPH LAVENBURG

2

In this wind-up alarm clock (below), the alarm system (blue) is powered by the same mainspring (B) that powers the rest of the clock. The alarm has its own operating wheel, called the barrel (M), attached to the bottom of the mainspring. The barrel turns only when the alarm is ringing. As it turns, the mainspring unwinds. To see how the alarm works, look at the drawing at right.

In the clock, the alarm wheel (1) and the hour wheel (2) fit under the dial. Follow the yellow cone to their exact position. Note that the hour wheel has two slots, and the alarm wheel has two tabs sticking down. Now, read below.

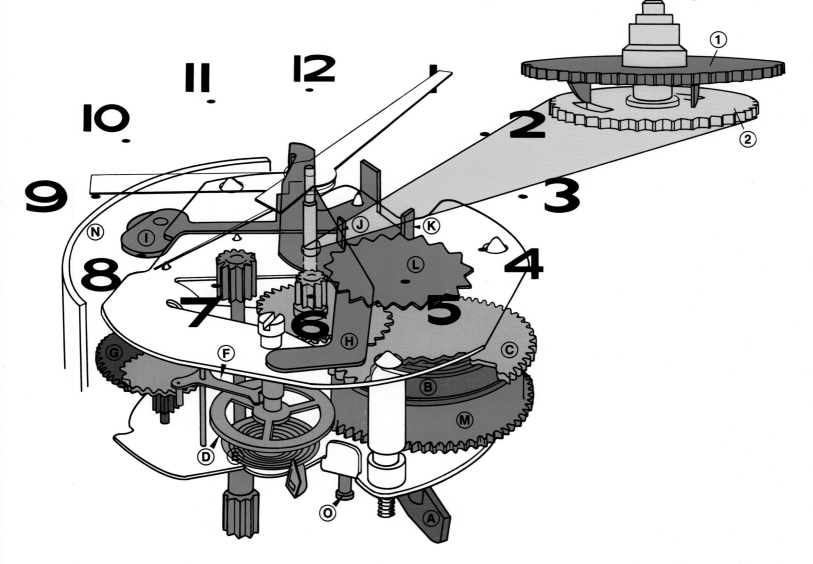

When you move the alarm hand to 8 o'clock on the dial, the alarm wheel also moves. It lines up at its 8 o'clock position (above). When the hour hand reaches 8 o'clock, the slots in the hour wheel line up with the tabs on the alarm wheel (top right). As the tabs slip into the slots, the hour wheel springs up against the alarm wheel. This action releases an arm (H) that was pressing down on the alarm hammer (I). The alarm hammer is now free to move. At the other end of the hammer stem are two more tabs (J and K). These tabs fit into the teeth of the alarm escape wheel (L). Another wheel connects this wheel to the barrel (M), on the mainspring. Freeing the hammer also frees the alarm escape wheel and the barrel. This makes the mainspring unwind, turning the two wheels. The teeth of the escape wheel hit the tabs on the hammer stem. This makes the hammer swing to and fro and strike the metal gong (N), causing the ringing sound you hear. You shut off the alarm by pushing in the alarm control button (O).

MARVIN J. FRYER

13

How a Toaster Works

It seems so simple. Drop in the bread. Push the handle down. Wait a few moments. Then—*pop!*—out comes hot, golden brown toast.

In a way, a toaster works much the same way a light bulb does. Inside both the bulb and the toaster, electric current flows through a wire, causing the wire to heat up. In the light bulb, the heated wire provides light. In the toaster, the heated wire browns a piece of bread. The toaster, of course, does more. It also turns itself off and delivers the finished toast.

Early toasters had no automatic timers. They simply cooked the bread. Then, in 1918, a mechanic named Charles Strite added a "clockwork" timer that would turn off the toaster when the bread had cooked a certain length of time. Manufacturers later added devices that popped up the toast. Now, toasters have become even more complex machines.

The most complicated part of any toaster is the timer. Toaster timers rely on a basic principle: Heat makes things expand. Heat is a form of energy. Energy heats up atoms, the tiny particles that make up all things. Heated atoms are more "energetic" than cool ones. They need more room, just as you do when you feel energetic. When an electric current heats up certain metal parts in the timer of a toaster, they expand and take up more room.

Not all metals expand at the same rate, however. The key part of the timer is a bar made of two different metals. As the bar heats, one of the metals expands faster than the other, causing the bar to bend toward a lever that shuts off the electric current. When the current turns off, the metal bar begins to cool. It loses heat and begins to contract, or draw together. As it returns to its original shape, it hits another lever. This lever pops the toast up, all warm and ready to eat.

On the next pages, you can look inside a toaster to see how this metal bar—and the other parts of a toaster—work.

Clean and simple, a shiny toaster case (left) hides a complex oven inside. The main operating handle at the top of the slot and the dark/light knob below it turn the toaster on and tell it how long to cook. To see what happens when you turn the knob and push down the handle, read on.

N.G.S. PHOTOGRAPHER JOSEPH LAVENBURG

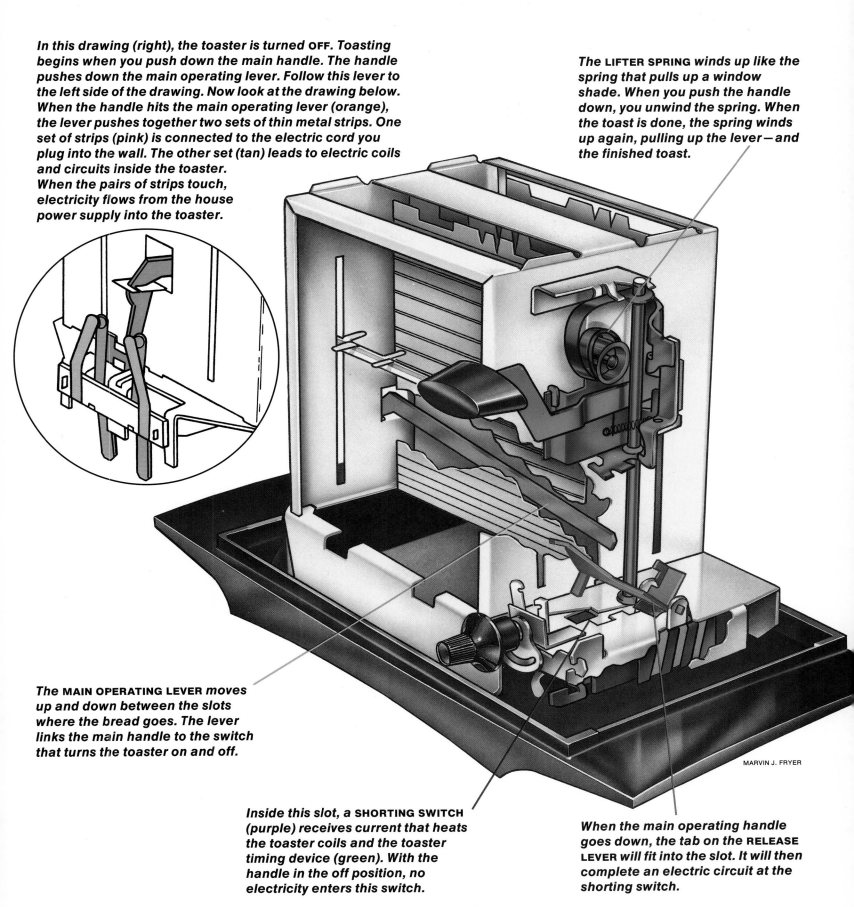

In this drawing (right), the toaster is turned OFF. Toasting begins when you push down the main handle. The handle pushes down the main operating lever. Follow this lever to the left side of the drawing. Now look at the drawing below. When the handle hits the main operating lever (orange), the lever pushes together two sets of thin metal strips. One set of strips (pink) is connected to the electric cord you plug into the wall. The other set (tan) leads to electric coils and circuits inside the toaster. When the pairs of strips touch, electricity flows from the house power supply into the toaster.

The LIFTER SPRING winds up like the spring that pulls up a window shade. When you push the handle down, you unwind the spring. When the toast is done, the spring winds up again, pulling up the lever—and the finished toast.

The MAIN OPERATING LEVER moves up and down between the slots where the bread goes. The lever links the main handle to the switch that turns the toaster on and off.

Inside this slot, a SHORTING SWITCH (purple) receives current that heats the toaster coils and the toaster timing device (green). With the handle in the off position, no electricity enters this switch.

When the main operating handle goes down, the tab on the RELEASE LEVER will fit into the slot. It will then complete an electric circuit at the shorting switch.

MARVIN J. FRYER

In this drawing, the toaster is in the ON position. The main handle is down, and the main lever has completed an electric circuit, bringing current into the toaster.

The LIFTER SPRING (blue) has been pulled tight. When the toast is done, the spring will release, pulling the toast to the top of the toaster.

You pushed the MAIN HANDLE down after you put the bread in the toaster. When the toast is done, this handle will slide up by itself.

The DARK/LIGHT CONTROL KNOB controls the length of time the bread-toasting coils stay on. When you adjust the knob, you change the distance that the metal timing device must travel before it strikes a lever that frees the toast.

MARVIN J. FRYER

The heart of the timer is the BIMETAL BAR (green). When electric current enters the coils (red) wrapped around one end of the BIMETAL BAR, the other end of the bar expands and bends.

With the RELEASE LEVER (brown) down, current has entered the SHORTING SWITCH (purple) and flowed into the coils that toast the bread. The current also has flowed into wire coils wrapped around the end of the timing device.

At right, you can see the part of the toaster that controls the timing. It is positioned as it would be if you had just pushed down the handle that lowers the bread. The release lever is in position under the heat-up lever. In this position, the release lever opens a circuit at the shorting switch (purple), forcing the toaster current to flow through the bimetal bar and enter the coils that toast the bread. Soon these coils will turn red. Metals expand when they are heated. The bimetal bar consists of two metals locked together. One metal heats and cools faster than the other. This causes the end of the bimetal bar shown here on the left to bend as it is heated and then straighten out again as it cools. As it bends, two tabs (1 and 2) attached to the bimetal bar release first the heat-up lever and then the cool-off lever so that the toast can pop up. Below, you can see the action step by step.

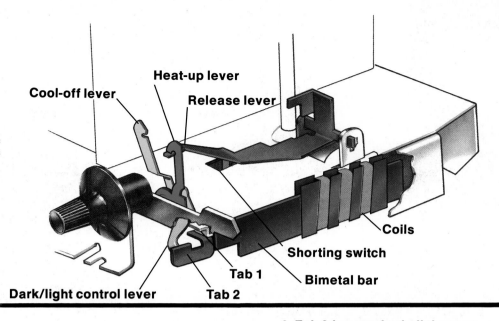

Cool-off lever
Heat-up lever
Release lever
Coils
Shorting switch
Tab 1
Bimetal bar
Dark/light control lever
Tab 2

1. Now you are looking at the same part of the toaster that you saw above, but from the front. The top of the heat-up lever (dark blue) has trapped the end of the release lever (brown), holding it down. The bimetal bar (green) has begun to heat up. When heated, it will slide slowly to the left. As tab 1 slides past the bottom of the cool-off lever (yellow), the cool-off lever slowly moves into an upright position.

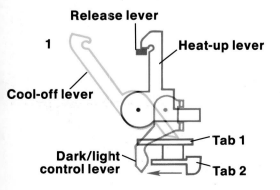

Release lever
1
Heat-up lever
Cool-off lever
Tab 1
Dark/light control lever
Tab 2

2. Now the cool-off lever is straight up and down. The bimetal bar is still heating, so it's still sliding to the left. Tab 2 is pushing back on the dark/light control lever attached to the heat-up lever.

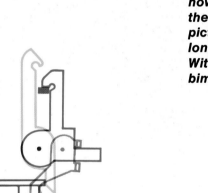

2

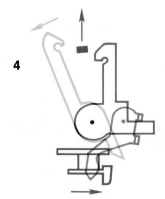

4

3. Tab 2 has pushed all the way back on the dark/light control lever. Now it pulls down on the back of the heat-up lever. This causes the top of the heat-up lever to tilt to the right. The release lever moves up until it is caught by the cool-off lever. By moving up, however, it has broken contact with the shorting switch (see large picture above), so current no longer flows into the bimetal bar. Without current to heat it, the bimetal bar will begin to cool.

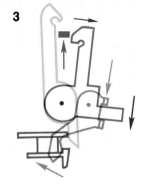

3

4. As it cools, the bimetal bar moves back to the right. When it moves far enough to the right, tab 1 will push the bottom of the cool-off lever to the right again, too. That will make the top of the cool-off lever tilt to the left and free the release lever. When this happens, the toast will pop up.

How a Compass Works

When you walk through your neighborhood, familiar sights and street names guide you. But how can you find your way in a strange place? If you know the direction in which you want to travel, a magnetic compass will help you. This kind of compass uses the earth's magnetism to indicate directions.

People usually think of a magnet as a small piece of metal that attracts iron. Yet the earth itself behaves as if it were a huge magnet. Like every magnet, the earth has two areas where its magnetic pull is strongest. These are the north and south magnetic poles.

Don't confuse the magnetic poles with the true, or geographic, North and South Poles. The geographic poles mark the ends of the earth's axis—an imaginary line about which the earth rotates. On a globe, lines of longitude extend from one geographic pole to the other. Mapmakers draw in these lines to help locate places.

As with any magnet, an invisible magnetic field exists between the earth's magnetic poles. This field is usually represented by lines that show its location and its strength. The magnetic field is strongest where the lines are closest together—at the magnetic poles.

It is the earth's magnetic field that makes a compass work. A compass has a magnetized steel needle. The needle rotates freely on a pin called a pivot. Opposite poles of two magnets attract. The field around the earth's north magnetic pole attracts the needle's south magnetic pole. Therefore, this end of the needle, often painted red, points north.

On the next page, you will find out how to use a compass. Remember that the compass needle points to magnetic north. By following it, you will go in a northerly direction, but not true north.

Near their home in Riverton, Wyoming, Jon Larson teaches Tanna Frye, 7, and Robbie Burleson, 13, how to use a map and a compass. Their map, known as a topographical (top-uh-GRAF-uh-kul) map, has curved lines. The lines show the contours, or shapes, and the elevations of every hill and valley around them. By studying the map, they can figure out where they are and how to get where they want to go. Their compass will keep them heading in the right direction.

DAVID P. JOHNSON, N.G.S. STAFF (BOTH)

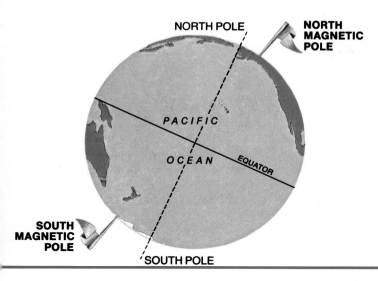

NORTH POLE

NORTH MAGNETIC POLE

PACIFIC

OCEAN

EQUATOR

SOUTH MAGNETIC POLE

SOUTH POLE

The earth's magnetic poles

If you look at a globe (left), you can easily find the North Pole and the South Pole, also known as the geographic poles. These spots mark the ends of the earth's axis — an imaginary line about which the earth spins. If you look more closely, you will find another set of poles. They are the magnetic poles, the areas where the pull of the earth's magnetism is strongest. Each magnetic pole lies within 2,000 miles (3,219 km) of its comparable geographic pole. The magnetic poles move from time to time, but the pull of the field remains northerly. It is the magnetic field between these poles that attracts a compass needle.

ROBERT W. NORTHROP

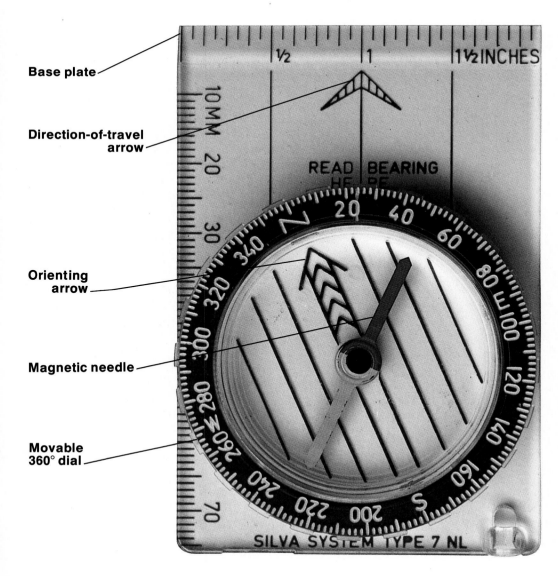

Base plate

Direction-of-travel arrow

Orienting arrow

Magnetic needle

Movable 360° dial

READ BEARING HERE

SILVA SYSTEM TYPE 7 NL

Finding your way

This compass (left) is called an orienteering (or-ee-en-TEER-ing) compass. Like most good camping compasses, it has a freely spinning magnetic needle, a transparent base, and a movable dial. Marked on the dial are N, S, E, and W, for north, south, east, and west. You can also see numbers around the dial. These stand for degrees, units of measurement for a circle. A circle has 360 degrees. The needle of the compass moves in a clear liquid. This liquid helps bring the needle to a quick stop. Operating a compass takes practice. Suppose you want to find a cabin that you know is east of where you are standing. Here is how to find your way there. First, hold the compass so the direction-of-travel arrow on the base plate points directly ahead of you. Then twist the dial until the E lines up with the direction-of-travel arrow. Now, keeping the compass level, turn your body until the red end of the magnetic needle lines up with the orienting arrow on the dial. The direction-of-travel arrow on the base now faces east, 90 degrees from north. You head in the direction that this arrow points.

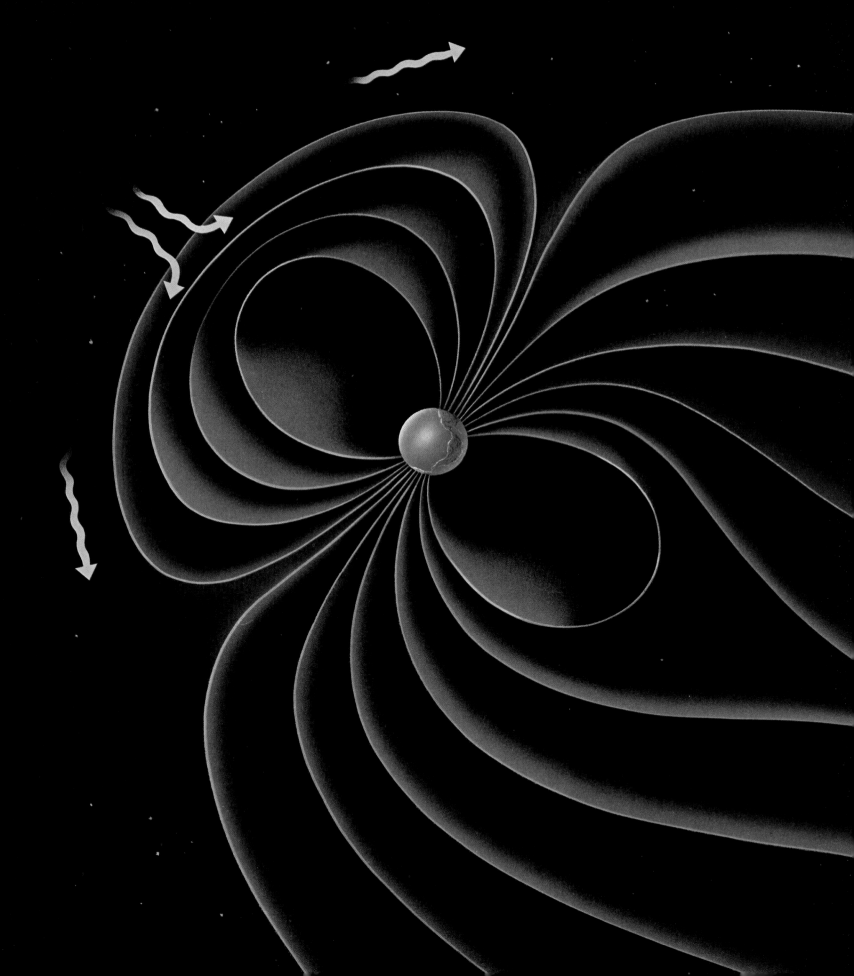

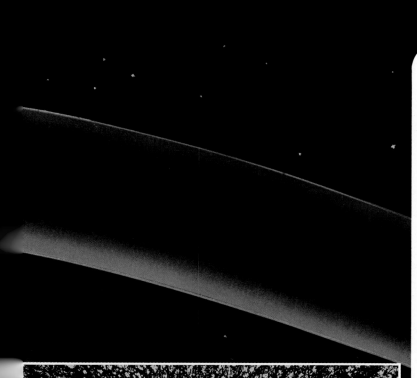

The earth's magnetic field

The earth acts as if a giant bar magnet were buried inside it between the North Pole and the South Pole. An invisible magnetic field forms a shield around the earth. Maps and charts often show this field as lines.

Most scientists believe that the field is created by the movement of liquid metal circulating deep within the earth. Although you cannot see the field, certain instruments inside satellites have mapped its shape and location. The field protects you from deadly electrically charged particles that flow from the sun. These particles form what scientists call the solar wind. The charged particles in the solar wind could destroy all life on earth. Fortunately, the earth's magnetic field traps them or forces them to bounce off into space.

You have seen how magnetism can help you by controlling a compass and by shielding you from dangerous charged particles. Magnets perform other valuable services as well. Generators use magnets to help produce most of the electricity that people use every day in the home, at work, and at school. Magnets are also used to change electric signals into sound waves in telephones, television sets, and radios. Without magnets, these devices would be silent. Magnetism combined with electricity also allows electric motors to work, as you will see on the following pages.

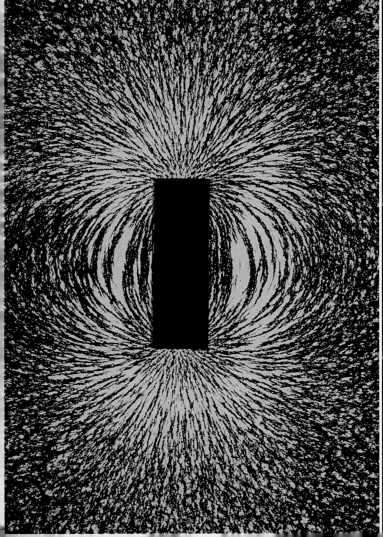

The earth's magnetic field is invisible. If you could see it, however, it would look like the large painting on these pages. Magnetic energy comes out of the core of the earth, extends out from the north magnetic pole, and loops back to enter the earth again at the south magnetic pole. In the painting, lines represent the field. The field is strongest where the lines are closest together, in the regions of the magnetic poles. The field spreads out and grows weaker away from the poles. The yellow arrows represent the solar wind—a flow of electrically charged particles from the sun. The solar wind presses against those lines of energy that face the sun. As the wind sweeps past the earth, it tugs some of the field away with it.

MARVIN J. FRYER

An easy way to see the force of a natural field in action is to ask your teacher to help you set up an experiment like this one (left). Place a piece of thin paper on top of a bar magnet. You should be able to see the magnet through the paper. Sprinkle iron filings onto the paper. The filings will line up with the bar's magnetic field, with the heaviest concentration at the poles. The earth's magnetic field would look much the same way if there were no solar wind to change or distort its shape.

N.G.S. PHOTOGRAPHER JOSEPH H. BAILEY (INSET)

How Electric Motors Work

Every day, you use objects that are powered by electric motors. When you vacuum your room or blow-dry your hair, an electric motor usually does the work. Do you know what makes such a motor work? It's a force known as electromagnetism.

Hans Christian Oersted, a Danish scientist, helped make the electric motor possible. In 1820, Oersted demonstrated an important natural law (below). He placed a magnetic compass under a wire. The wire ran north and south, so the compass needle and the wire were parallel. Then Oersted connected the ends of the wire to opposite poles of a large battery. When electricity from the battery flowed through the wire, the compass needle rotated until it pointed west. When Oersted reversed the direction of the current in the wire, the needle rotated until it pointed east. When he turned off the current, the needle again pointed north.

Oersted reasoned from this that the electric current had created a magnetic field around the wire. The field had pushed the north-seeking end of the compass needle away from magnetic north.

Oersted had proved that electricity and magnetism are related. His wire had become an electromagnet. An electromagnet is a temporary magnet—one that is a magnet only when electricity flows through it. A current-carrying loop of wire forms a magnetic field similar to that formed by a permanent bar magnet. Each has a north and a south pole. Poles that are unlike attract and poles that are alike repel. In an electromagnet, however, the poles change places when the direction of the current reverses.

Most modern electric motors contain electromagnets made of wires wound into coils. Coils make very powerful electromagnets; when current flows in one direction through a coil, the magnetic field around one winding strengthens the field around the next winding of the coil. Wrapping the coil around a core of iron or steel strengthens the magnetism in the coil even more, because the metal core also becomes magnetized.

On these pages, you'll see how two sets of coils create the energy that makes an electric drill work.

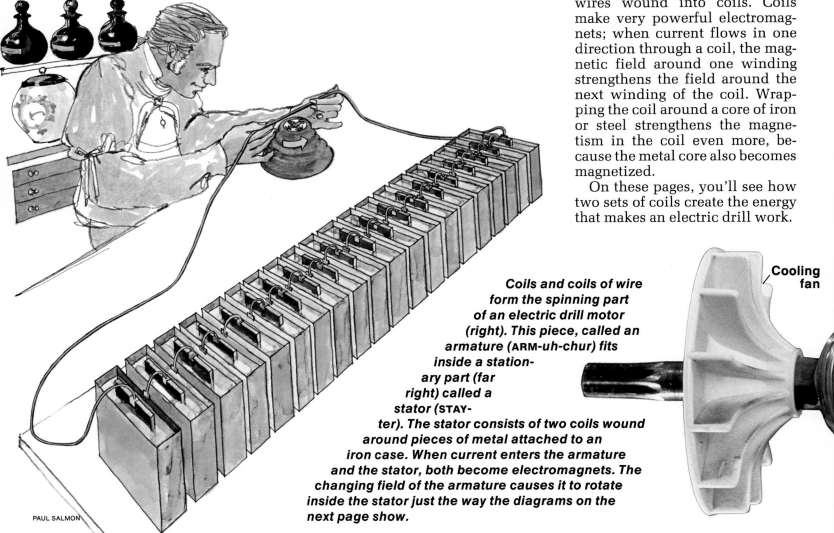

PAUL SALMON

Coils and coils of wire form the spinning part of an electric drill motor (right). This piece, called an armature (ARM-uh-chur) fits inside a stationary part (far right) called a stator (STAY-ter). The stator consists of two coils wound around pieces of metal attached to an iron case. When current enters the armature and the stator, both become electromagnets. The changing field of the armature causes it to rotate inside the stator just the way the diagrams on the next page show.

Cooling fan

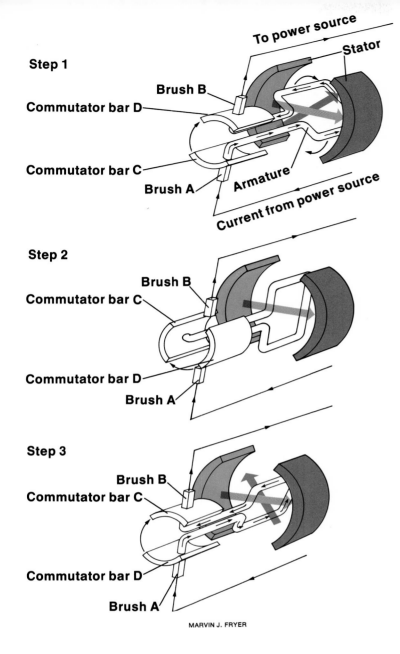

Step 1

To power source

Stator

Brush B

Commutator bar D

Commutator bar C

Brush A

Armature

Current from power source

Step 2

Brush B

Commutator bar C

Commutator bar D

Brush A

Step 3

Brush B

Commutator bar C

Commutator bar D

Brush A

MARVIN J. FRYER

Suspend a bar magnet by a string. Then bring another bar magnet near it. The suspended magnet will rotate until its poles line up with the opposite poles of the magnet you're holding. This same rotating motion drives an electric motor. In the simplified drawings at left, you can see how one motor—a direct-current motor—works. The motor contains two electromagnets: an armature and a stator. The armature rotates freely inside the stator. The stator does not move. When you switch on the motor (step 1), current flows into the stator coils and creates a north (blue) and a south (red) magnetic pole in the stator. The stator coils have been removed from the drawing, but you can see them in the photograph below. A magnetic field flowing from north to south (brown arrow) forms between the stator poles. The current enters and leaves the armature only when two pieces of carbon called brushes (A and B) touch two metal plates called commutator bars (C and D), as they do here. Current in the armature creates another magnetic field (green arrow). The north pole is at the point of the arrow. Unlike poles attract; the S pole of the stator (red) pulls on the N pole of the armature. So the armature rotates clockwise. But as this happens, the commutator bars rotate past the brushes (step 2). Without current from the brushes, the armature loses its magnetic field (the green arrow disappears). Momentum keeps the armature rotating, however, until it again comes into contact with the brushes and current again flows through the armature (step 3). But now, the commutator bars are in the opposite positions from the ones in step 1, and the current—and the armature's magnetic field—reverse. The south pole of the armature's field (bottom of the green arrow) is now next to the south pole (red) of the stator and the north pole (top of the green arrow) is next to the north pole (blue) of the stator. Like poles repel, so the armature keeps spinning. These steps keep repeating as long as the current is on.

N.G.S. PHOTOGRAPHER JOSEPH LAVENBURG (BOTH)

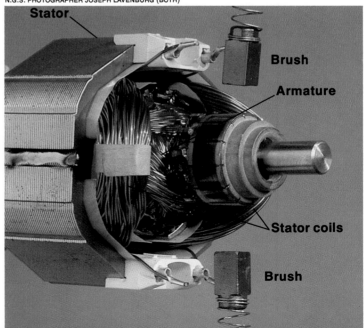

Stator

Brush

Armature

Stator coils

Brush

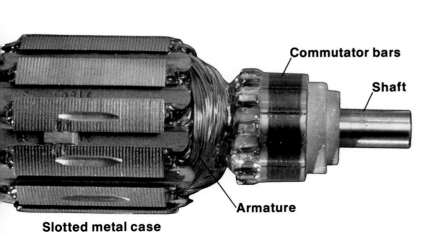

Commutator bars

Shaft

Armature

Slotted metal case

How Vending Machines Work

Have you ever put a coin in a chewing gum machine or bought a can of soda from a soft-drink machine? If you have, you know what a vending machine is. It's a kind of robot salesperson. It automatically gives out candy, drinks, and other items when you put in the right coins.

The first people to use a vending machine may have lived in ancient Egypt. According to legend, in about the second century B.C. there existed in a temple in Alexandria a device that dispensed "holy" water.

Vending machines appeared in the United States in the 1890s. People could buy stamps and buttons from them. In the early 1900s, manufacturers developed machines for selling gum, peanuts, and other products. These machines all worked with levers and gears.

Then manufacturers added electricity. This gave the machines lighted signs to catch the customer's eye. It gave them refrigeration units to keep foods fresh and drinks cold. It gave them heating units to make hot coffee, tea, and soup. And it gave them the ability to do arithmetic. Today, many vending machines add up nickels, dimes, and quarters, and give back change.

All these improvements have made vending machines more complicated than before. But the machines still have just two basic jobs to do. First, they identify and count the inserted coins. Then they deliver the product that the customer selects.

Vending machines do break down sometimes. However, if a vending machine swallows your coin without giving you what you ordered, the machine itself probably isn't at fault. It's likely that there is a bent coin somewhere inside, jamming the works.

Gum-ball machines like the one at left work simply. The balls of chewing gum rest on a flat metal plate with several holes in it. Gum balls fall into the holes. Putting a penny in the slot frees the lever so you can slide it several inches. Sliding the lever moves the plate around until one gum ball lines up with a channel beneath it. Then that gum ball falls into the channel and rolls down into your hand. Modern vending machines (below and right) can produce a meal a minute.

DAVID FORBERT/SHOSTAL ASSOCIATES, INC. (LEFT)

MARTIN ROGERS (BELOW AND OPPOSITE, ALL)

Chips? Cookies? Crackers? This vending machine (left) offers you a choice of 25 kinds of snacks. Larger items fill the top shelves. Smaller items line the lower shelves. Circular rings called augers (AWE-gers) hold the snacks in place. An auger is shaped like a metal spring. One or two snacks fit inside each coil. When you insert coins and select a snack, the auger holding the snack makes a complete turn for a large snack or a half turn for a smaller one. The snack then falls from the coil into a tray behind the panel marked PUSH. As an auger delivers a snack, it brings another snack from behind to replace it.

Adding things up

A closeup view of the door panel of a snack machine (below) shows food-selection buttons, a coin slot, and a coin-return button. Directly behind this panel are systems that direct some of the machine's operation. One tests inserted coins to make sure they are real. It also adds them up. Then it signals a dispensing unit to deliver the selected item. To find out what happens when you insert a coin and push a button, turn the page.

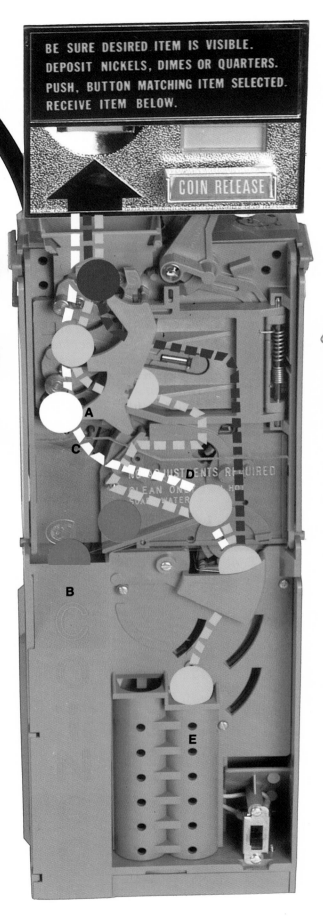

BE SURE DESIRED ITEM IS VISIBLE.
DEPOSIT NICKELS, DIMES OR QUARTERS.
PUSH, BUTTON MATCHING ITEM SELECTED.
RECEIVE ITEM BELOW.

COIN RELEASE

Real coin or fake? The coin-acceptor unit (left) can tell. When you insert a dime (white path), it falls first into a cradle (A). Here, it is checked for size. If it fails the test, it drops down into the coin-return channel (B). A dime that passes the test rolls down a ramp into a slot that checks its weight and thickness (C). Then it rolls on to a serration detector (D). This is a thin wire that catches the ridges on the edge of a dime and stops it. The dime then drops down a ramp into a collection box (not shown) at the bottom of the unit (not shown). A smooth-edged fake (red) would keep on rolling until it hit a wall and bounced back into the return channel. Nickels (yellow path) and quarters (green path) also go through tests. Nickels drop into change-making tubes (E) or into the collection box.

Directly behind the coin-acceptor unit is an electrical panel (above). The panel makes the coin acceptor work. Electric current enters the panel at top left. A transformer (A) then reduces the current to the amount that the machine actually needs. Electric switches called vend line relays (B) turn on motors that deliver selected items. Resistors (C) protect the relays from voltage spikes— sudden increases in current. When the power is off, magnets (D) block the coin paths. Any coins dropped in fall into the return slot. Switches (E) enable an attendant to set the prices for each row of products.

Making the delivery

A wire auger — a large coil — makes a half turn and a package of crackers falls (left). The crackers will land in a tray, where a customer can pick them up. These augers hold double rows of products. A small electric motor (above) operates each auger. The motor turns the coil. The coil does not move forward. It rotates in place. As it does, other packages automatically move forward and fill the space left when a snack is sold. Thicker augers on upper shelves hold larger products. They make a full turn when delivering the products they hold.

The sign on the milk-vending machine below would attract attention on a hot day. This refrigerated unit sells small cartons of whole milk, skim milk, and chocolate milk. Other types of refrigerated machines sell soft drinks in paper or plastic cups. The machines make the drinks by mixing flavored syrups with cold water and ice. For fizzy drinks, the machines add carbon dioxide to the water.

MARTIN ROGERS (ALL)

7

How a Camera Works

A camera is an instrument that captures images from life by directing reflected light onto film. Everything about a camera has to do with the control of light. If too much light strikes the film, the image will be overexposed and look washed out. If too little light reaches the film, the image will be underexposed and look dark.

Every camera has four basic parts: the body, the film holder, the shutter, and the lens. The body is a light-proof box that supports the other parts. The film holder keeps the film in place at the back of the camera. The shutter opens to let in light. The lens gathers the light and focuses it onto the film.

There are many different types of cameras. The one on these pages is a 35-mm SLR. The "35-mm" part of the name means the film is 35 millimeters wide. "SLR" means single-lens reflex. Photographers like this type of camera because it is easy to handle and because a mirror-and-prism system inside enables them to see almost the same image that the lens "sees." Also, the 35-mm SLR camera can be fitted with a variety of lenses—normal, closeup, zoom, wide-angle, and telephoto. Different lenses help you take the exact picture you want. Often, the camera has a built-in exposure meter that shows when you have set the camera's controls so the lens lets in the proper amount of light.

Cameras of one form or another have been in use for centuries. Within the last hundred years, advances in photography have put cameras in nearly everyone's hands. A big breakthrough was roll film. Another was the development of a camera small enough to be carried anywhere and to take pictures without the use of a supporting stand.

Photographing in color became popular along with the 35-mm SLR. Color film is sensitive to three colors of light—red, green,

and blue. During development, dyes are formed that produce all the other shades.

One of the first things you should learn about a camera is how to load it with film. With some cameras, you just drop in a film cartridge. The 35-mm SLR camera is a little more complicated. The film comes rolled up in a cassette. A strip of film sticks out of the cassette. To load the camera, open the back of it. Drop the cassette into the pocket under the film rewind lever. The drawing on the next page will help you find the pocket. Now pull the strip of film across the shutter and tuck it into a slot in the take-up spool. Advance the film until the holes in the edges of the film fit over the sprocket teeth. Then close the camera and advance the film to number 1 on the frame counter.

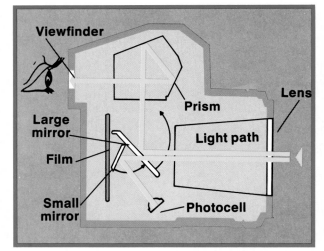

This diagram (above) shows how light moves through a 35-mm SLR (single-lens reflex) camera with a built-in exposure meter. Light enters through the lens. Some of the light hits a small mirror that aims it onto a photocell. This cell measures the amount of light entering the camera. The light also bounces off a large mirror and through a prism—a glass crystal—which focuses it into the viewfinder.

DRAYTON HAWKINS, N.G.S. STAFF

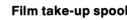

Film take-up spool

The **SHUTTER RELEASE BUTTON** *does two things. It pulls the large and small mirrors out of the path of light entering the lens. At the same time, it opens the shutter, allowing a measured amount of light to strike the film.*

The **SHUTTER SPEED DIAL** *helps control the amount of light that strikes the film. It does this by presetting the amount of time the shutter stays open. The dial goes from 1 (one second) to 1000 ($^1/_{1000}$ of a second). The dial often contains an inner dial (yellow). You use this dial to set the camera to the film speed. The speed is printed on the film package beside the letters ASA (American Standards Association).*

The **FILM REWIND LEVER** *ends in a fork that fits into the film cassette. When you've exposed all the film, you press and hold the film release button on the bottom of the camera. This allows the sprocket roller to turn backward. Now you flip up the center piece of the rewind lever and wind the film completely back into the cassette.*

Film advance lever

Frame counter

Inside story

This is what you would see if you could look through the front of a 35-mm SLR (single-lens reflex) camera body with the lens removed. The artist added colors to identify the different systems that work together to take a picture. These include the film advance (blue), the shutter (orange), the exposure meter (yellow), and the film rewind (green).

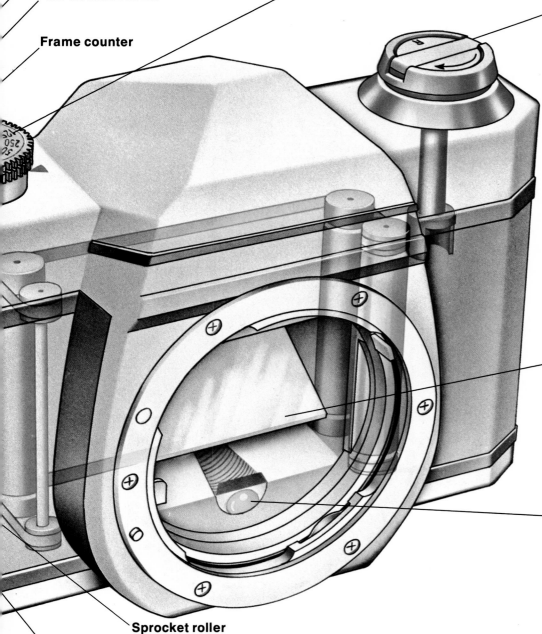

The **LARGE MIRROR** *reflects most of the light entering through the lens to the photographer's eye. In this camera, a smaller mirror behind the large mirror reflects light onto the photocell.*

The **PHOTOCELL** *measures the amount of light entering the camera. It relays this information electronically to a system in the viewfinder that shows you when you have adjusted the camera to get a properly exposed picture.*

Sprocket roller

Film release button

MARVIN J. FRYER

The shutter

You are looking through the back of a 35-mm SLR camera body (middle picture). The shutter consists of two lightproof curtains. The curtains are attached to two sets of rollers. When you press the shutter release button to take a picture, one curtain snaps from one roller onto the other, in the direction of the arrows. The second curtain follows the first one. This leaves a gap between the curtains. The film is exposed by light coming through the gap. The amount of light that comes through depends on the width of the gap and on its speed as it moves across the film. The number you select on the shutter speed dial controls this action. The faster the shutter speed, the smaller the amount of light that reaches the film.

The SHUTTER SPEED DIAL (right) is located next to the film advance lever. If you want to shoot a scene such as the stars at night, set the shutter speed dial at B. The shutter then stays open as long as you hold down the shutter release button. After each picture, you advance the film and reset the shutter by pushing the lever with your thumb.

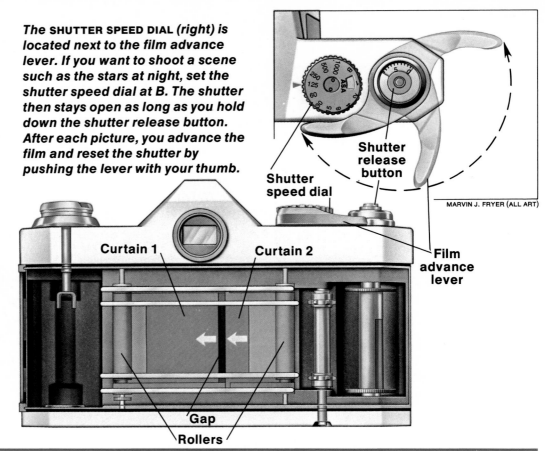

MARVIN J. FRYER (ALL ART)

Curtain 1   Curtain 2

Shutter speed dial   Shutter release button   Film advance lever

Gap

Rollers

How the shutter controls motion

A fast shutter speed freezes action. The picture below left, for example, was shot at $^1/_{500}$ of a second. Everything in the picture is sharp and clear—bicycle, rider, background. In fact, there is very little sensation of motion. It almost looks as though Deborah Hurwitz, 12, of Bethesda, Maryland, is balancing on her bicycle in one spot. Now look at the action in the other picture below. The background is blurred. But Deborah and the bicycle are still quite sharp. If the photographer had simply used a slow shutter speed, the girl and bike would be blurred and the background sharp. But he used a slow shutter speed, $^1/_{60}$ of a second, and "panned." Panning means following the action with your camera just as you would with your eyes.

Experienced photographers use many different kinds of lenses. The beginner needs only one — a 50-mm lens (below). This is the lens that usually comes with the camera. A lens is made up of several pieces of shaped glass. Between the pieces of glass is a diaphragm (DIE-uh-fram). This is a set of curved, overlapping metal plates. The diaphragm can be set to let in a little light, or a lot of light. The opening in the diaphragm is called the aperture (AP-er-chure). Aperture sizes are measured in f-stops. The seven diaphragm drawings below show that aperture sizes in a 50-mm lens often range from f/2, the biggest opening, to f/22, the smallest. The bigger the opening, of course, the more light that comes through. Each f-stop after f/2 lets in half as much light as the stop ahead of it. F-stops can be confusing because the numbers get larger as the opening gets smaller. To keep the numbers straight, think of the f-number as the bottom of a fraction. Put a 1 where the f is. Then it's easy to remember that $\frac{1}{22}$ is smaller than $\frac{1}{2}$, for example. F-stops control how much light gets through the lens. In addition to determining the

N.G.S. PHOTOGRAPHER JOSEPH H. BAILEY (ABOVE, AND OPPOSITE)

exposure of the photograph, the f-stops determine how much of the picture, from foreground to background, will be in focus. This is called depth of field. The smaller the opening, the deeper the depth of field. Proof? Note the depth of field difference in the pictures above. The one on the left was shot at f/22. The one on the right was shot at f/4. Look inside your viewfinder for the exposure meter. The needle will balance in the middle when you select the correct f-stop for (1) the amount of light on your subject, (2) your shutter speed, and (3) the type of film you are using.

f/22 f/16 f/11 f/8 f/5.6 f/4 f/2

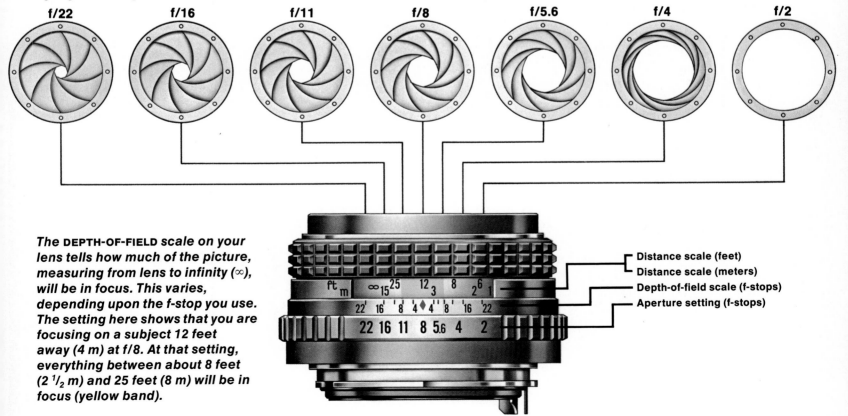

The DEPTH-OF-FIELD scale on your lens tells how much of the picture, measuring from lens to infinity (∞), will be in focus. This varies, depending upon the f-stop you use. The setting here shows that you are focusing on a subject 12 feet away (4 m) at f/8. At that setting, everything between about 8 feet (2 $\frac{1}{2}$ m) and 25 feet (8 m) will be in focus (yellow band).

Distance scale (feet)
Distance scale (meters)
Depth-of-field scale (f-stops)
Aperture setting (f-stops)

31

8

HOW NEON WORKS

What a gas! You see it almost every time you walk down a city street. It's in signs that advertise movies. It's in signs that point the way to fast-food spots. It's in signs that warn NO LEFT TURN. What is the gas that makes the letters and lines in all these signs stand out? It's neon—one of six gases called "rare gases" because they make up such a small portion of the atmosphere.

A French chemist named Georges Claude helped make these bright signs possible. Early in this century, Claude found a way to take the rare gases out of the atmosphere. He put them in glass tubes and used them for lighting.

A mixture of gases makes up our atmosphere. The lightning that brightens the sky is nothing more than a huge electric discharge, or spark, streaking through these gases. An electric spark also streaks through a neon tube, but the neon gas gives the spark an orange-red color. Other rare gases produce different colors. The gas argon, for example, creates pale lavender light. To make different shades, craftsmen may use neon and argon together or with another substance, such as mercury. The craftsmen may use a tinted tube or a tube coated with a combination of powders that give off a colored glow when energy from the gas inside hits them. In these ways, as many as 40 bright colors can be produced.

Craftsmen bend the glass tubing into hundreds of shapes. They can make words, directional arrows, cowboy hats, even dancing dogs.

A neon sign is economical to operate. It uses a lot of energy to get started, but little once it begins to glow—and it will keep on glowing for many years.

Neon signs brighten night skies in much of the world. Some signs let people know where they can buy "donuts" or have their fortunes told. Others decorate homes and art galleries. Some of these signs contain gases other than neon. But people commonly call all signs that work this way after the original neon signs.

SING

neon

GYPSY
TEA
KETTLE

TEA LEAVES READ GRATIS
TABLE

heaven

Steaks

HOW
THINGS
WORK

AAA

Bar-B-Q To Go

OSKAR'S
ICE CREAM
CONE

In a shop in Landover, Maryland, glassblower Don Grieser twists a heated glass tube (above). He's forming it into the letter "O" for a sign. He heated the tube over a gas burner until it was flexible enough to form into a rounded shape. He works with his bare hands so that he can feel the temperature of the glass. Years of experience have taught him the exact moment a tube is hot enough to shape.

COTTON COULSON (ABOVE AND OPPOSITE, BOTH)

Very carefully, Grieser prepares to seal an electrode onto a tube (right). He will attach an electrode to each end. The electrodes, which are simply rods of solid iron, will carry electricity into and out of the tube. After fitting the two pieces of glass together, Grieser will roll them in the flame until they melt together and form an airtight seal.

CHARLES E. HERRON, N.G.S. STAFF

Grieser uses a device called a spark coil to create a spark in invisible gas sealed inside a glass container (right). The gas glows orange-red. The color shows that this gas is neon. Only neon gas gives off such a color. Grieser will pump the air from an electrode-fitted tube and replace it with neon from this container.

READY, SET, GLOW!

To find out how neon glows, look at the drawing below. Like all things, neon is made up of atoms. An atom consists of a nucleus and electrons. The nucleus has a positive electric charge; the electrons have a negative electric charge. Electrons circle around the nucleus in different-size orbits, depending on how much energy they have. A neon atom has ten electrons. In the drawing, the neon atom has been simplified to look like this: ⊖⊕ . The ⊖ represents an electron in a neon atom; the ⊕ represents the nucleus. The gas also contains free, or unattached, electrons.

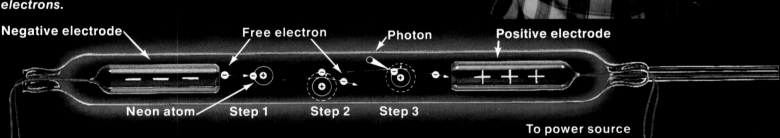

Negative electrode   Free electron   Photon   Positive electrode

Neon atom   Step 1   Step 2   Step 3

To power source

When electric current enters the electrodes at either end of the tube, the electrodes take turns being positive and negative as the current from the power source changes direction. Opposite electric charges attract, much the same way that opposite poles of magnets do. So the free electrons immediately speed toward the positive electrode (1). Now watch what happens when one of those fast-moving free electrons hits an electron in orbit in a neon atom. The orbiting electron does not receive enough energy from the collision to knock it away from the atom. But it does jump into a wider orbit (2). A fraction of a second later, however, it falls back into its normal orbit (3). As it does, it gives off the energy it received during the collision in the form of a tiny burst of light called a photon. The energy of the photon determines the color of the light. In neon, the color is orange-red. Billions of photons give the tube an even, continuous glow.

MARVIN J. FRYER

Putting the finishing touches on his creation, Grieser uses blockout paint to cover the parts of the glass tubing that he doesn't want to glow (right). Two words light up as electricity disturbs the pure neon gas inside the tube. A third word will be added to form the title of this book. Look at the cover or page 33 to see the finished sign.

How Lasers Work

Laser light can do things no other kind of light can do. Its energy is so concentrated that it can accurately measure the thickness of a gnat's antenna—or the distance to the moon. In the hands of a surgeon, it can mend a damaged inner eye painlessly.

Laser light can do all these things—and more—because it is the most coherent, or organized, light source known. All light consists of tiny particles of energy called photons (FOE-tahns). In laser light, these photons are all of the same wavelength, or color. Light is made up of waves that have high points called crests and low points called troughs. The distance between crests is called the wavelength. In laser light, the crests and troughs of each wave line up with the crests and troughs of the waves around it. This is called being "in phase," or in step—like members of a marching band. It is this coherence that makes a laser beam shoot straight ahead in one narrow, powerful beam.

The devices that produce laser light come in many sizes. Some are as small as a grain of sand. Others are nearly as long as a football field. Some laser beams—those that scan price labels at supermarket checkout counters, for example—give off less light than an ordinary flashlight. They are not dangerous to use. But high-powered lasers, such as those used to cut metals, can cause serious injury. People who use them must wear special equipment for protection.

The first laser to use visible light was the ruby laser, developed in 1960. It consisted of three basic parts: 1) A fluorescent (flur-ESS-uhnt) material, in this case chromium atoms, inside a rod of artificial ruby crystal. A fluorescent material is one that will give off light when it absorbs energy from another source, such as other light. 2) A source of energy, in this case a flashtube. 3) Two mirrors reflecting inward, one at either end of the rod. One mirror is designed to reflect only partially. Lasers still have these basic parts.

Today's lasers often use fluorescent substances, such as those contained in certain gases, liquids, and other types of crystals. The color of the beam changes with the wavelength given off by the fluorescent substance. All lasers work in the same basic way, however.

These superlights face a bright future. New uses are constantly being found for them. Dentists, for example, have discovered that a weak beam "washed" over the teeth seems to prevent decay. Lasers may someday transmit most electronic communications. A single beam could carry billions of telephone conversations at the same time—and that's not just a lot of talk!

At a supermarket in Alexandria, Virginia, a laser helps cashier Paula Sims speed customer Annis Vaughn through the checkout line. Mrs. Sims holds a can that has a series of black lines printed on the label. As she passes the can over a window in the counter, a laser beam scans the lines. The beam reflects off the lines and spaces and bounces back to a light-sensitive plate. It translates the light into an electric signal, which flashes to the store's computer. The computer matches the signal with the product description and price. Instantly, this information appears on a display on the cash register.

N.G.S. PHOTOGRAPHER JOSEPH H. BAILEY (ALL EXCEPT BOTTOM CENTER)

To operate the laser, Monroe flips a power switch near the top of the cane (below) with his thumb. The high beam shoots through the window just below the switch. The middle and low beams come out at different angles from the larger window underneath. Special batteries power the laser beams.

Guiding light

A laser cane for the blind helps Charles Monroe, of Arlington, Virginia, avoid obstacles in his path (above). The cane contains three lasers that send out invisible beams (below). The beams detect obstacles at different heights and distances. One beam points upward (line A). If a tree branch or other high object blocks the way, the beam bounces back (dotted line) to a receiver in the cane. The receiver is a light-sensitive cell that converts the light energy into an electric signal. That signal triggers a high-pitched alarm. The second beam (line B) detects obstacles at waist level. The alarm it produces has a medium pitch. The third beam, aimed at the ground, works somewhat differently. It sets off an alarm only when it does __not__ bounce back to the receiver. An irregularity at ground level — a curb, say, or a drop-off — sends the beam off at an odd angle. That triggers a low-pitched alarm. Then the user knows to step cautiously.

5½ feet (1½ m)

A
B
C

◄— 3 feet (1 m) —►
◄———— 12 feet (3½ m) ————►

ADAPTED BY DRAYTON HAWKINS, N.G.S. STAFF, FROM ART; COURTESY OF NURION, INC., WAYNE, PA.

Near the middle of the cane, three light-sensitive cells, called photocells (below), sense the reflected beams. Monroe can control the volume of the alarms. The middle beam triggers a vibration in addition to the alarm. If he chooses, he can turn off the sound and use only the vibration.

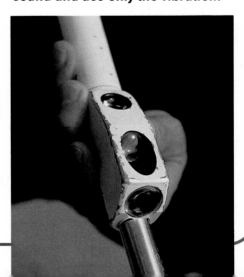

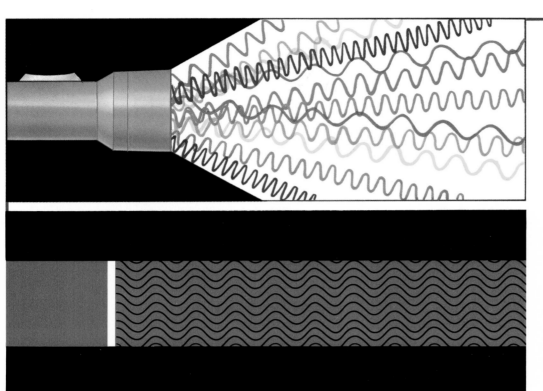

Organizing light

Laser light moves in waves, just as ordinary light does. But laser light waves differ from waves of ordinary light in two important ways. The waves of energy that make up a flashlight beam (left, top), for example, are of many different wavelengths, or colors. These waves travel in different directions. The beam they produce spreads out like a fan. That causes the light to fade and soon disappear. Light waves from a laser, on the other hand, are coherent. That is, all the laser light beams are the same wavelength, and they all travel in the same direction (left, bottom). Laser waves travel together uniformly, in parallel paths. Coherence keeps laser light from spreading out as it travels.

MARVIN J. FRYER

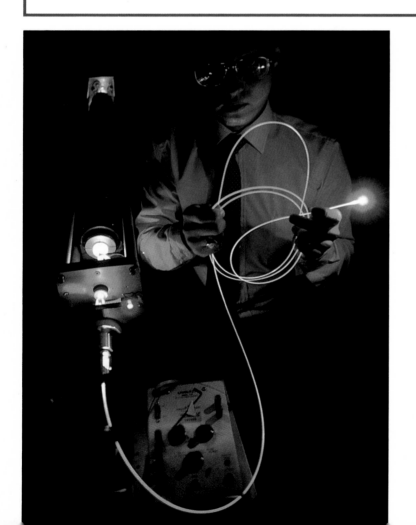

In a device used to perform delicate surgery, a laser beam travels through a flexible glass fiber (left). Dr. Terry Fuller, a biophysicist with Sinai Hospital, in Detroit, Michigan, shows how the fiber enables surgeons to bend and guide the beam through the body. The box in front controls the strength of the beam. The beam flows through the glass fiber much as water flows through a hose. It ends in a pinpoint of light. The light can quickly and safely seal off tiny, broken blood vessels inside the body. Dr. Fuller developed this laser device in response to a request by surgeons. They wanted a way to control internal bleeding. Laser beams have other medical uses as well. Laser light can be focused so exactly that it can be used effectively in the brain, in the eye, and in the inner ear. Depending on its wavelength, a laser beam can perform bloodless surgery, mend torn tissue, and sizzle away unwanted growths, all without damaging neighboring healthy tissue.

HANK MORGAN/RAINBOW

Laser beams atop the Empire State Building, in New York City (right), pierce the night sky. The city put on this red and blue laser show in 1981 to celebrate the building's 50th birthday. On clear nights, people could see the brilliant beams from 10 miles (16 km) away.

ANDY LEVIN/BLACK STAR

39

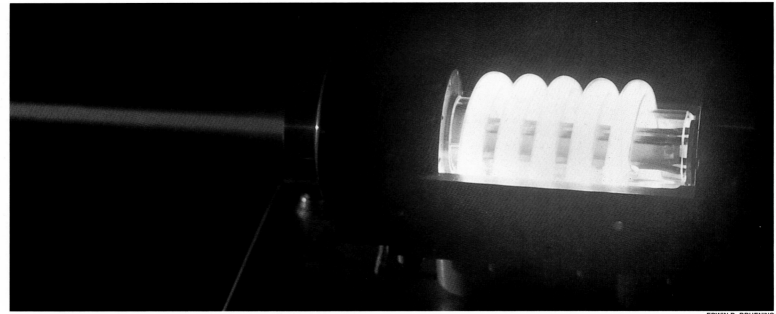

EDWIN B. BRUENING

A jet of pure red light shoots out of a ruby laser. Like all lasers, this laser consists of three basic parts. First, it has a fluorescent (flur-ESS-uhnt) substance. Fluorescent substances give off light when struck by certain kinds of energy. This laser uses an artificial ruby rod, which contains fluorescent chromium atoms. Second, it has an energy source—here, a coiled flashtube—that pumps light energy into the fluorescent substance. Third, it has a pair of mirrors. These cap the ends of the rod. Below, you'll see how these three parts create laser light.

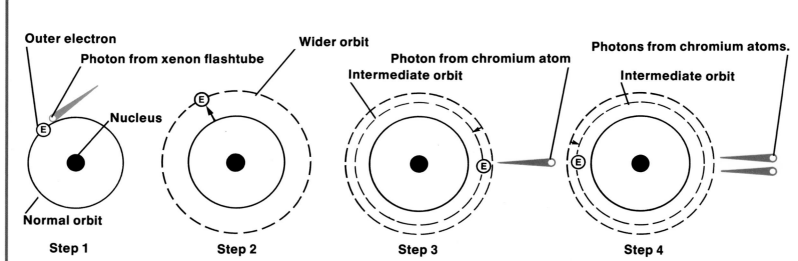

Outer electron
Photon from xenon flashtube
Wider orbit
Nucleus
Normal orbit
Step 1

E

Step 2

Photon from chromium atom
Intermediate orbit
E
Step 3

Photons from chromium atoms.
Intermediate orbit
E
Step 4

Trillions of chromium atoms form the fluorescent substance in a ruby laser. The drawings above show a chromium atom greatly simplified. Like every atom, it has a nucleus, or central part. Other particles called electrons orbit the nucleus at different levels, or distances. In a ruby laser, the flash from a flashtube filled with atoms of xenon (ZEE-nahn) gas sends out tiny light particles called photons (FOE-tahns). When a photon from a xenon atom (green) hits an outer electron in a chromium atom (1), it "excites" the electron—gives it energy. The electron jumps into a wider orbit (2). Almost immediately, however, the electron returns to an intermediate, or lower, orbit (3). From there, the electron releases the extra energy in the form of a photon (red). If this photon from the chromium atom hits another chromium electron that is in an intermediate orbit, that atom will release a photon (4). Both photons leave the atom in the same direction, with their waves in step. It is this second collision that makes laser light coherent.

MARVIN J. FRYER (ABOVE, AND OPPOSITE, ALL)

Building a beam

A laser produces enormous numbers of well-organized photons and releases them in a tightly packed beam. In a ruby laser, the process begins when the xenon-filled flashtube gives off a flash of light. Photons from the xenon atoms fly about like sparks from Fourth of July sparklers. You can see them in the cutaway diagram at left. The artist painted them as green streaks. You can see the actual color of xenon in the photograph (opposite). Some photons from the xenon tube hit chromium atoms in the ruby rod. This causes the chromium atoms to release red photons. The metal case bounces outward-bound photons back into the rod. Two mirrors, one at each end of the rod, reflect the red photons moving inside the rod so they bounce back and forth from one end to the other. By design, the mirror at the left end of the rod reflects totally. The mirror on the right reflects slightly less.

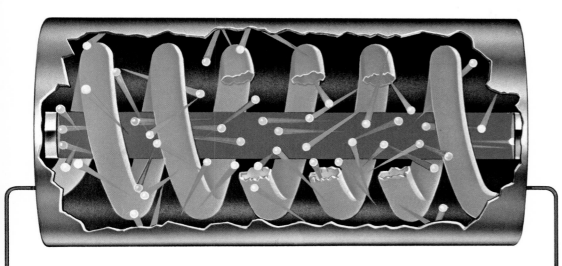

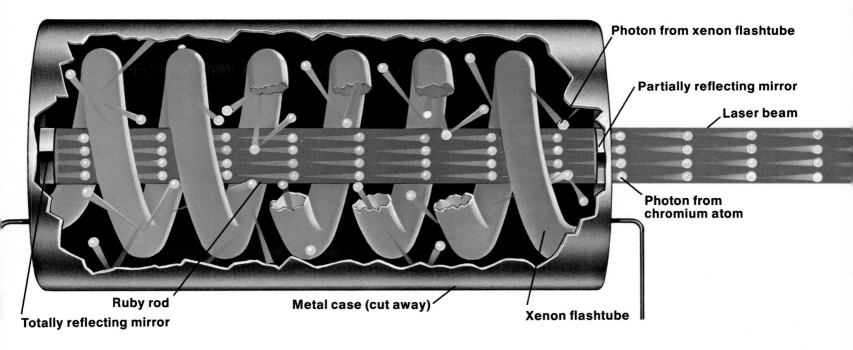

Photon from xenon flashtube

Partially reflecting mirror

Laser beam

Photon from chromium atom

Ruby rod

Metal case (cut away)

Xenon flashtube

Totally reflecting mirror

Bouncing back and forth between the two mirrors, the red photons multiply as they collide with excited chromium atoms. Soon the photons are all moving through the center of the rod in step with each other. In a few thousandths of a second, the number of red photons builds up to a high intensity and they escape through the weaker mirror in a brilliant beam of laser light (above), the most intense light source known.

10

How Clarinets and Pianos Work

If you've ever blown across the top of an empty soft-drink bottle, you know that your breath can make the air inside the bottle give off a sound. Air moving inside a clarinet makes a sound, too. You can hear the sound. What you can't do is see what is happening to the air.

Air, like every substance, is made up of atoms and molecules—groups of atoms. When you speak, your vocal cords cause the molecules in the air to begin to vibrate, or move back and forth. They can vibrate as fast as thousands of times a second. These vibrations are what make sound. As the sound spreads out, vibrating air molecules bump into air molecules next to your ears. These molecules set your eardrums vibrating, too. This is how you hear sound.

Vibrating air inside a clarinet and a bottle produces sound in the form of musical notes. The notes may be high or low. The highness or lowness of a note is called its pitch. Pitch is determined by the frequency, or number of vibrations a second, of the waves producing it. A column of air in an empty bottle generally produces a low-pitched note. What happens if you add some water to the bottle and blow? The column of air is now shorter, and the pitch is higher. The shorter the air column is, the higher the note is.

To get different notes out of a clarinet, you don't pour water into it, of course. Holes in the sides of the instrument do the job. Pads cover many of the holes. If you blow into a clarinet when these holes are covered, the air travels out the end of the clarinet. You produce the lowest-pitched note possible. If you open some holes, most of the air escapes before it reaches the end. This air produces higher-pitched notes, because the length of the vibrating air column is shorter.

When you blow across the bottle, your breath makes the air inside move. In a clarinet, you don't move the air directly. You blow past a thin, flat reed attached to the mouthpiece. The reed vibrates as you blow, making the air inside the clarinet vibrate, too.

Tester Nancy Walters (left) knows how a clarinet should sound. At the Selmer Company, in Elkhart, Indiana, she checks newly finished clarinets for air leaks, nicks, and other problems. Gauges aid in her inspection.

N.G.S. PHOTOGRAPHER JOSEPH H. BAILEY

If you were to slice a clarinet mouthpiece in half, here's what you would see (below). The flat reed almost covers an opening in the back of the mouthpiece tube. A clarinetist holds the end of the mouthpiece—shown here in a circle—between the lips. The reed rests on the lower lip. When the player blows into the instrument, forcing a stream of air into the mouthpiece, the reed's tip vibrates. This causes the air in the tube to vibrate. The vibrations create a musical sound.

A thin, tapered reed (above) gives a clarinet its particular sound. This kind of instrument has a single reed that fits inside the mouthpiece. Similar instruments, such as the oboe, have a double reed. A reed can be made of cane or of plastic. Feathery thin at the tip, the reed bends easily. Some clarinetists shave the reed even thinner to make it more flexible. Metal strips called the ligature hold the reed in place.

Low notes and higher ones

A. Sound wave in a long air column results in a low-pitched note.

B. Sound wave in a short air column results in a high-pitched note.

The shape of a clarinet makes air blown into it form a column. When the reed sets the air column in motion, you hear a sound. The shorter the air column is, the higher pitched its sound is. A player changes the length of the air column by covering and uncovering holes. These drawings show air columns moving through a simplified clarinet. Curved lines represent moving air. In picture A, air travels a long way down the tube. Because all of the holes are blocked, the air travels through the tube and leaves at the other end. This produces a low-pitched note. In picture B, three holes are uncovered, so most of the air will escape near the mouthpiece. This shortens the column and produces a higher-pitched note.

Pulling a lever, a technician lowers a drill at the Selmer factory (above). He is drilling a hole in the wooden tube that will form the body of the clarinet. A special machine clamped to the tube moves it a certain distance down a track and turns it when necessary. This helps the technician drill the holes at just the right places. So that it will create notes of the correct pitch, a clarinet must have holes placed at exact points along the length of the tube.

Ready for playing, a clarinet lies on an open music book (left). Round pads cover certain holes. A player controls the pads by pushing keys. Metal rings surround other holes. By pressing a ring with a finger, a player can cover a hole. Clarinets are available in several different pitches, depending on the model.

Music from strings

When Herb Ostroff sits down at the piano in a New York City music store and begins to play, his hands move back and forth across the keyboard, his fingers striking the keys. He can play loud or soft, slow or fast, high or low—whatever the music and his mood call for. As he plays, thousands of piano parts respond to his touch. Together, all these parts produce the sounds he wants.

Simply put, a piano is a large sound box. Strung tightly inside the box are many metal strings. Eighty-eight hammers, all in a row, face the strings. When a player presses a key on the piano keyboard, the connecting hammer hits the proper string or set of strings. As each string vibrates, the air around it vibrates, too. When the vibrations reach your ears, you hear a sound.

Pianos come in two shapes. In a grand piano, such as the one shown below, the strings lie flat. In a vertical piano, the strings run up and down. Strings vary in length and in thickness. The shorter and thinner the string is, the higher-pitched the sound is. Strings range in length from about 2 to 87.5 inches (5-222 cm).

By itself, a vibrating string does not produce much sound. You can test this easily. Tie a piece of cord to a chair, stretch the cord tight, and pluck it. You'll see the cord vibrate, and you'll hear a soft sound. The thin cord moves only a small amount of air and so produces weak sound waves as it vibrates. Now hold the free end of the cord against a windowpane and pluck the cord. The vibrating cord makes the glass vibrate, too. The glass moves many more air molecules, and this helps the cord make a louder sound.

The secret of a piano's strong sound lies in a large, thin piece of wood. It is found at the bottom of a grand piano and at the back of a vertical one. It is called the soundboard. As the piano strings vibrate, the soundboard vibrates as well, giving the waves more energy. In pianos, strings produce the sounds, but soundboards give them enough energy to make your eardrums vibrate.

Onlookers gather as Herb Ostroff plays a grand piano at the Baldwin Piano & Organ Company store in New York City. Thick wood forms the sides of the piano case. About 9,000 piano parts, including wires, pins, levers, and springs, help produce the sounds. A thin wooden soundboard inside the case amplifies the sound—makes it louder. Ostroff can change the sound by striking the keys sharply or gently, and by pushing pedals with his feet.

45

❶

To create piano notes, you need metal strings, something to make the strings vibrate, and something to keep them quiet when they are not in use. The system that performs these jobs is called the action. The picture below shows a demonstration model of the action of a vertical piano. A piano key (A) sets the action in motion. When you press down on one end of the key (not shown), the end you see here rises like a seesaw. It strikes the wooden rocker arm (B). One end of the rocker arm (the left end in this picture) tilts downward. It moves a metal rod attached to the damper (C). The damper is made of soft material called felt. When the damper rests against the string, it keeps the string quiet. Moving the damper allows the string to vibrate. As the damper moves away from the string, levers and springs attached to the other end of the rocker arm go into action. They throw the hammer (D) against the string. The hammer strikes the string, causing it to vibrate, and then bounces back into place. As long as you hold your finger on a piano key, the string remains free to vibrate. When you remove your finger from the key, the damper falls back onto the string. The sound ends at once.

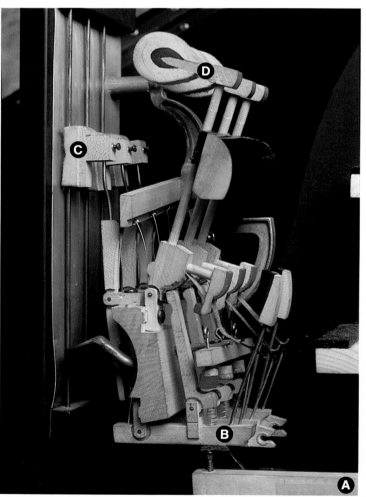

N.G.S. PHOTOGRAPHER JOSEPH H. BAILEY (ALL)

❷

A heavy cast-iron frame forms a piano's backbone. In a typical piano, about 230 strings stretch tight across the frame. In a grand piano, the strings pull on the frame with a combined pressure of at least 35,000 pounds a square inch (2,461 kg/cm²). Without an extra-sturdy frame, a piano would bend under the tension.

❸

On a typical keyboard, 88 piano keys join forces with 88 hammers to create 88 different notes. Each key operates the way a seesaw does. When you push down on one end, the other end rises. The way you press a key determines the kind of sound the piano makes. Pressing hard makes the hammer hit hard. This makes a loud sound. A light tap on a key makes a soft sound.

4

The soundboard, which enlarges the piano's sound, lies beneath the frame of a grand piano.

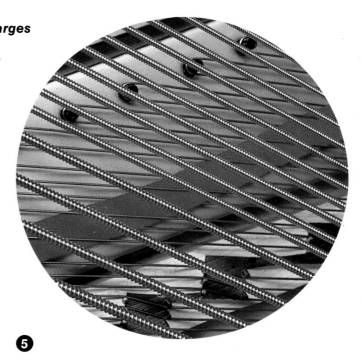

2

4

5

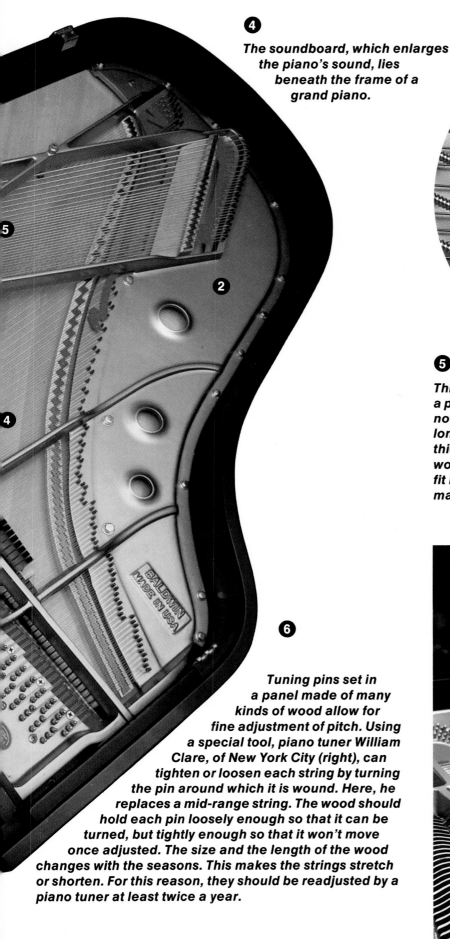

5

Thick and thin strings form a crisscross pattern inside a piano. The short, thin strings produce the piano's highest notes. To produce lower notes, the strings must be made longer or thicker—or both. If all strings were the same thickness, the lowest sounding string in a typical piano would have to stretch 21 feet (6 m). Such a piano wouldn't fit into many living rooms. To solve this problem, piano makers thicken some strings by wrapping them in wire.

6

Tuning pins set in a panel made of many kinds of wood allow for fine adjustment of pitch. Using a special tool, piano tuner William Clare, of New York City (right), can tighten or loosen each string by turning the pin around which it is wound. Here, he replaces a mid-range string. The wood should hold each pin loosely enough so that it can be turned, but tightly enough so that it won't move once adjusted. The size and the length of the wood changes with the seasons. This makes the strings stretch or shorten. For this reason, they should be readjusted by a piano tuner at least twice a year.

11

How a Stereo Radio Works

You slip the headset on, adjust the foam pads over your ears, flip a switch, and twirl a few dials. Suddenly, you're standing in the middle of an orchestra. Music is all around you. There, in your own backyard, you're attending a concert taking place a hundred miles away.

It is brought to you by a small box called a radio.

Radios—actually radio receivers—are so familiar today that it's difficult to believe that the first radio signals were sent less than a hundred years ago. In 1901, an Italian inventor named Guglielmo Marconi broadcast tapping signals by means of an invention he called a wireless. Marconi tapped out three clicks—the letter S in Morse code.

Soon, radio transmitters could broadcast signals from voices and music. The first radio receivers were about the size of large TV sets. Over time, the receivers became smaller and smaller. Today, people listen to radios small enough to tuck into a pocket. Many such radios produce life-like stereophonic sound.

Although radios have changed in size and shape, the way radio equipment works has not. Radios depend on radio waves transmitted from a broadcast antenna. These waves can travel thousands of miles in a second.

On these pages and the next, you'll find out how radio stations use radio waves to send programs from their studios to your ears.

Taking a break from homework, Will Toft, 13, listens to a radio in a park near his home, in Arlington, Virginia. His radio produces music in stereophonic sound. That means it operates almost as two radios in one. The radio antenna, in this case the earphone wires, receives a radio wave from the air. Electric circuits inside the radio remove two electric signals from the radio wave, improve them, and send one to each earphone. The earphones translate the signals back into sound waves. Will keeps the volume in the low to medium range. Listening to loud music over a long period of time could damage his hearing.

N.G.S. PHOTOGRAPHER JOSEPH H. BAILEY

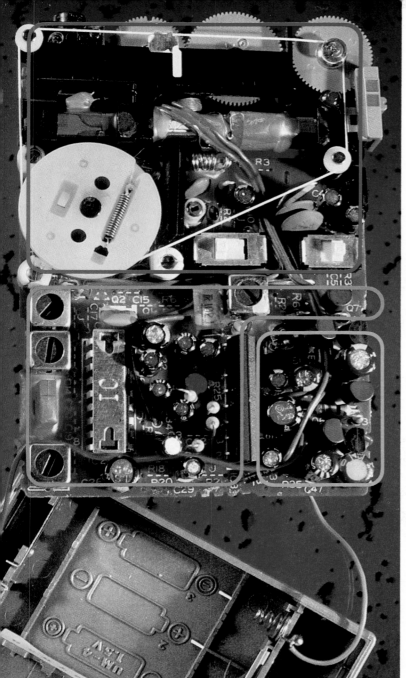

N.G.S. PHOTOGRAPHER JOSEPH LAVENBURG (ALL)

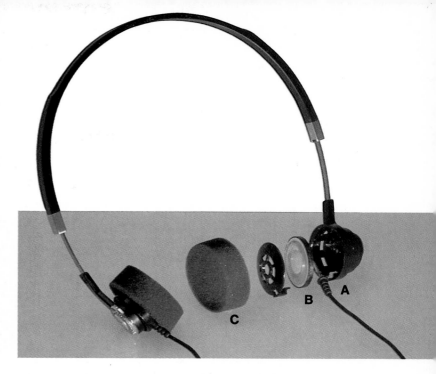

Earphones receive the electric signals from your radio and turn them back into sound waves. Above, you see the parts that make up each earphone. The housing assembly (A) protects the parts of the earphone. The speaker (B) contains a diaphragm (DIE-uh-fram), an extra-thin strip of flexible material. Changes in the strength of the electric signal set the diaphragm vibrating. These vibrations move the air to produce sounds that you can hear. A foam pad (C) makes the earphone comfortable to wear. To see how the signals reach your radio receiver, turn the page.

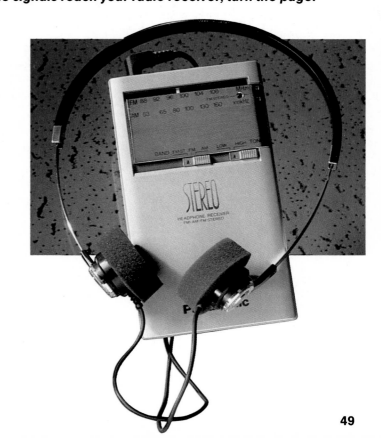

In radios like this one (right), many small parts work together to do a big job. They pull signals that you can't hear out of the air. Electric circuits inside the radio separate the signals and then change them into sound waves that you can hear. It takes three groups of electric circuits (above) to do this. The tuner (green) adjusts the receiver so it selects only the frequency of the station you want to receive. It rejects all the other frequencies hitting the antenna. The filters (orange) strip away a stronger radio carrier wave the station used to send the signal. The amplifier (blue) strengthens the signal again and sends it on to the speakers, which are the earphones of this radio.

From the station...

Inside radio station WHTW, two musicians step up to a pair of microphones (above). As they play, sound waves flow into the microphones. A magnet inside each microphone changes the sound waves into electric signals called audio-frequency (AF) signals (blue). A wire carries the two sets of AF signals through the station control room and into a transmitter. Inside the transmitter, a device called an oscillator (oss-uh-lay-tur) produces high-frequency radio waves, also called carrier waves (yellow). The oscillator combines each of the two AF signals with one carrier wave. The combined signals (green) now travel to a tower outside the station. At the top of the tower, a rod called a broadcast antenna sends the combined signals into the air in all directions.

AM or FM?

AF signals (1) that a microphone creates have a low frequency. That is, they make few vibrations per second. To broadcast these AF signals, the station transmitter mixes them with much faster and more powerful carrier waves (2). The transmitter does this by changing the carrier waves so they carry a code for the AF signal. This coding is called modulation. The transmitter can modulate the radio waves in one of two ways. It can change the amplitude, or distance from top to bottom, of the radio waves. This is called amplitude modulation, or AM transmission (3). The transmitter also can move the radio waves closer together or farther apart. This is called frequency modulation, or FM transmission (4). Whichever way the radio waves are changed, they now carry with them the coded AF signal.

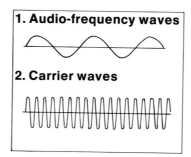
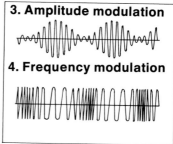

1. Audio-frequency waves

2. Carrier waves

3. Amplitude modulation

4. Frequency modulation

...to your ears

To receive your own personal concert from station WHTW, you turn on your radio and set the dial at that station's frequency (below). Turning the dial adjusts a tuner attached to the radio's antenna. The tuner makes the radio antenna receive only the signals from WHTW (green). The antenna rejects signals from all other stations. By now, the signals transmitted by WHTW have weakened. They may have traveled hundreds of miles before reaching you. An amplifier inside the radio strengthens the signals and sends them to a filter. The filter removes the carrier wave, leaving only the original AF signals (blue) that the microphones created at the radio station. On the last stage of their journey, the AF signals separate. Each signal now enters another amplifier, which strengthens it again. Then one signal enters the right speaker. The other signal enters the left speaker. The speakers act as microphones in reverse. When an AF signal enters a speaker, the signal causes a diaphragm inside to vibrate. The vibrations re-create the original sounds each musician made at station WHTW. Each speaker delivers a slightly different group of sounds, and the music sounds lifelike.

MARVIN J. FRYER (ALL)

Signals from many radio stations fill the air (above). If all these signals had the same frequency, voices and music from one station would interfere with broadcasts from other stations. To help prevent this from happening, the Federal Communications Commission (FCC), a government agency, assigns a certain frequency to each station. The station then generates radio signals on only that frequency. Some stations broadcast AM signals; others broadcast FM signals. AM signals travel farther than FM signals do, but FM signals are less affected by static electricity in the air than AM signals are.

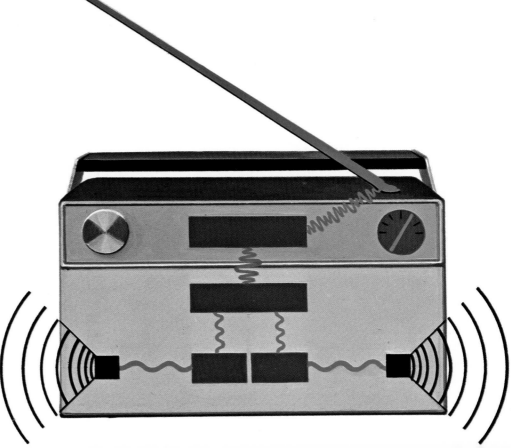

51

12

How Garage-door Openers Work

It's a rainy winter night. You've had a long drive, but finally you're home. Your mom drives the car into the driveway and stops. All of a sudden, you realize, "Oh, no! It's my turn to open the garage door." By the time you have finished, you're cold and wet, and your shin hurts because you stumbled over your bike in the dark.

An automatic garage-door opener would have turned on the light, opened the door, and closed the door—without a drop of rain hitting you.

Three forms of energy work together when an automatic garage-door opener operates: radio waves, electricity, and mechanical energy. Radio waves signal an electric motor to switch on. The motor turns a sprocket that pulls a chain. The chain moves the door up or down.

Many garage-door openers include several built-in safety features. But if you use an automatic opener, you should still take care to observe some rules:

● Never touch the transmitter when someone or something is near or under the door.
● Never try to run under a door while it is closing.
● Test the controls once a month to be sure they work smoothly.

Open up. A garage door opens automatically as a motorist sits comfortably in her car (right). She sets the door in motion by pressing a button on a small transmitter. The transmitter sends a coded signal to a receiver contained in a power unit inside the garage. The power unit then switches on a motor that pulls up the door. After the motorist drives into the garage, she can close the door behind her by pressing the button again.

ADAPTED BY PAUL SALMON FROM *HOW IT WORKS: ILLUSTRATED,* BY RUDOLF GRAF AND GEORGE WHALEN. COPYRIGHT © 1974 BY VAN NOSTRAND REINHOLD COMPANY.

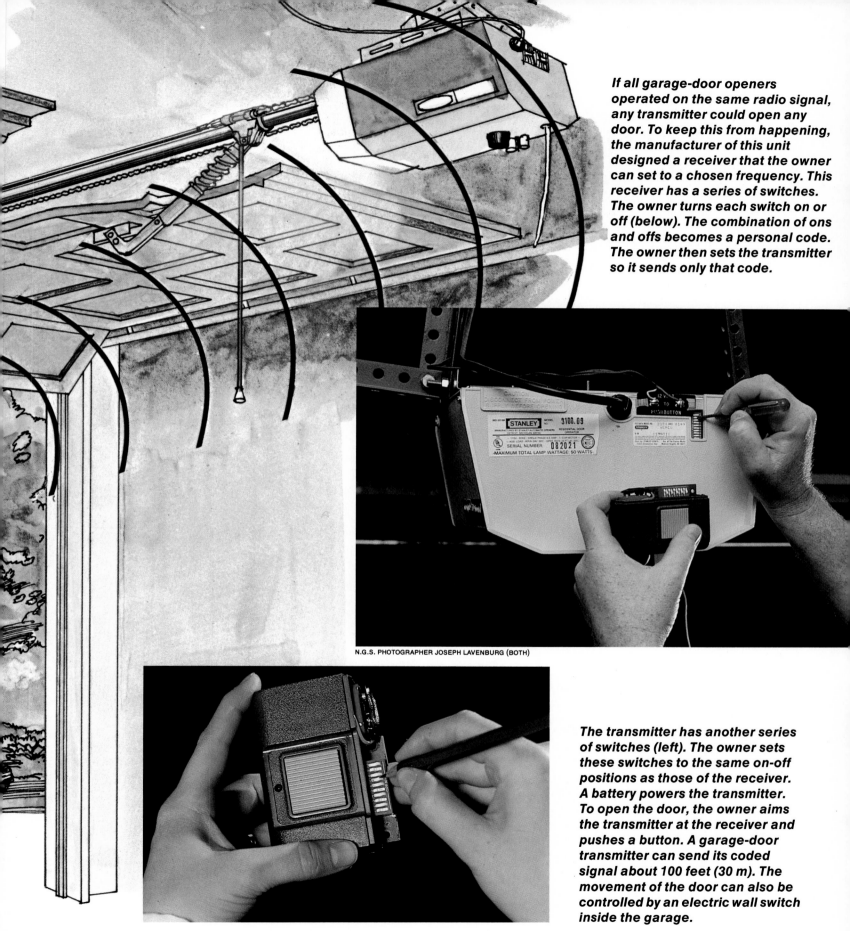

If all garage-door openers operated on the same radio signal, any transmitter could open any door. To keep this from happening, the manufacturer of this unit designed a receiver that the owner can set to a chosen frequency. This receiver has a series of switches. The owner turns each switch on or off (below). The combination of ons and offs becomes a personal code. The owner then sets the transmitter so it sends only that code.

N.G.S. PHOTOGRAPHER JOSEPH LAVENBURG (BOTH)

The transmitter has another series of switches (left). The owner sets these switches to the same on-off positions as those of the receiver. A battery powers the transmitter. To open the door, the owner aims the transmitter at the receiver and pushes a button. A garage-door transmitter can send its coded signal about 100 feet (30 m). The movement of the door can also be controlled by an electric wall switch inside the garage.

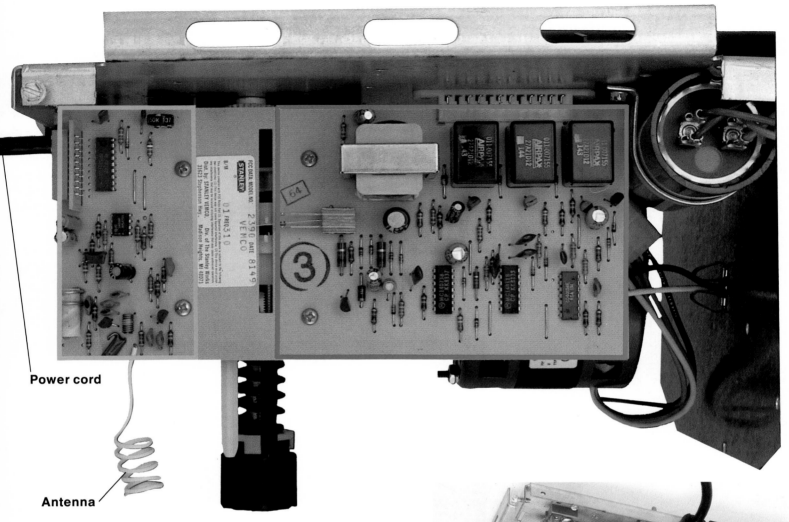

Power cord

Antenna

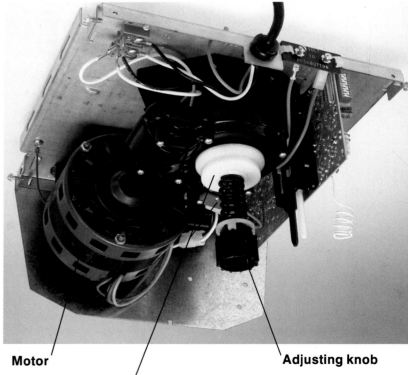

If you remove the cover from the power unit (above), you can see the "brains" of the door opener, the control board. The control board has a receiver section (orange) and a logic section (green). The antenna receives the signal from the transmitter and pulls it into the receiver section. There, the signal travels through circuits that stop the signal if it carries the wrong code. If it carries the correct code, the logic section switches on and figures out whether the door is open or closed. Then the logic section commands an electric motor to move the door in the opposite direction.

The "muscle" that makes the garage door move is a small electric motor (right) that turns a sprocket. The sprocket turns in one direction to open the door, and in the opposite direction to close it. If something gets in the way of the door while it is closing, a safety reverse mechanism sends the door up again. If the door hits something while opening, the reverse mechanism stops the door's movement. This reverse mechanism works automatically. The owner may adjust its pressure by turning a knob.

Motor

Adjusting knob

Safety reverse mechanism

N.G.S. PHOTOGRAPHER JOSEPH LAVENBURG (ALL)

This drawing shows a garage-door opener as seen from above. The electric motor inside the power unit turns a sprocket similar to those on bicycles. The sprocket pulls a continuous loop of chain and cable. Attached between the chain and the cable is a part called the traveler. The traveler slides back and forth on a metal tube that runs from the receiver to a bracket above the garage door. A J-shaped metal arm connects the traveler to the top of the garage door. As the chain-and-cable loop moves, the traveler either pulls the door up or pushes it down. If the electricity should go off, the owner can still open the door from the inside by pulling a cord hanging from the traveler. Pulling the cord temporarily disconnects the traveler from the chain-and-cable loop. The owner can then lift or lower the door by hand.

MARVIN J. FRYER

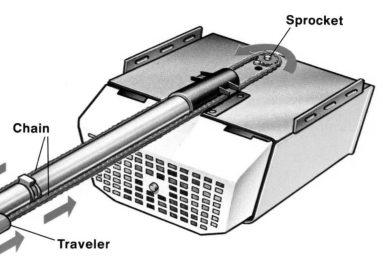

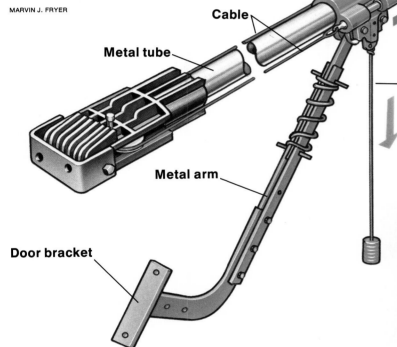

Sprocket

Chain

Cable

Traveler

Metal tube

Disconnect cord

Metal arm

Door bracket

As a garage door opens, a light flashes on automatically (right). When the driver pushed the button either on the transmitter or on the wall switch, the logic section of the control board completed a circuit, and the light went on. In some garage-door openers, a timer turns the light off after three or four minutes.

13

How Radar Works

A motorist in a hurry steps on the gas and exceeds the speed limit. As the car goes over a hill, a police officer steps from the shoulder of the road and flags down the car. The officer tells the driver: "You were going 45 miles an hour in a 25-mile-an-hour zone." The motorist groans, "Oh, no! It had to be radar."

Police officers do catch speeding motorists with the help of radar. Radar uses short waves called microwaves to "see" things not visible to the human eye. When microwaves strike an object, they bounce off. Some of them return to their source, just as a sound echo does. By studying the returning microwaves, people can find out the size and the location of objects hundreds—even thousands—of miles away.

Radar also can indicate almost instantly the speed and the direction of moving objects. Microwaves vibrate, just as sound waves and radio waves do. When microwaves bounce off a moving object, they vibrate differently when they return. If the object is moving away, the returning waves vibrate more slowly. If the object is approaching, the returning waves vibrate more quickly. Radar guns measure the difference between the outgoing and incoming vibrations. Then they translate that information into a number that tells how fast the object is moving.

Radar units come in many sizes and shapes. The radar guns used to detect speeding motorists fit in the hand. Other units are as large as buildings.

Scientists in several countries share the credit for developing radar. Work began in the early 1900s to find a way to use radio waves to detect enemy airplanes and ships. Today, radar stations along national borders can spot approaching missiles as far away as 3,000 miles (4,828 km).

Radar has other uses, too. Airplane pilots rely on it to avoid colliding with other planes and to land safely in bad weather. Mariners use it to navigate in coastal waters and to steer clear of icebergs. Weather forecasters track storms with radar. In the future, scientists may find even more uses for radar.

On the lookout for speeders, a police officer aims a radar gun at an approaching car and pulls the trigger. As he does, invisible microwaves bombard the car. Some of the waves hit the car and bounce back to the gun. Instantly, numbers showing the speed of the car appear in a display at the back of the gun.

N.G.S. PHOTOGRAPHER JOSEPH H. BAILEY

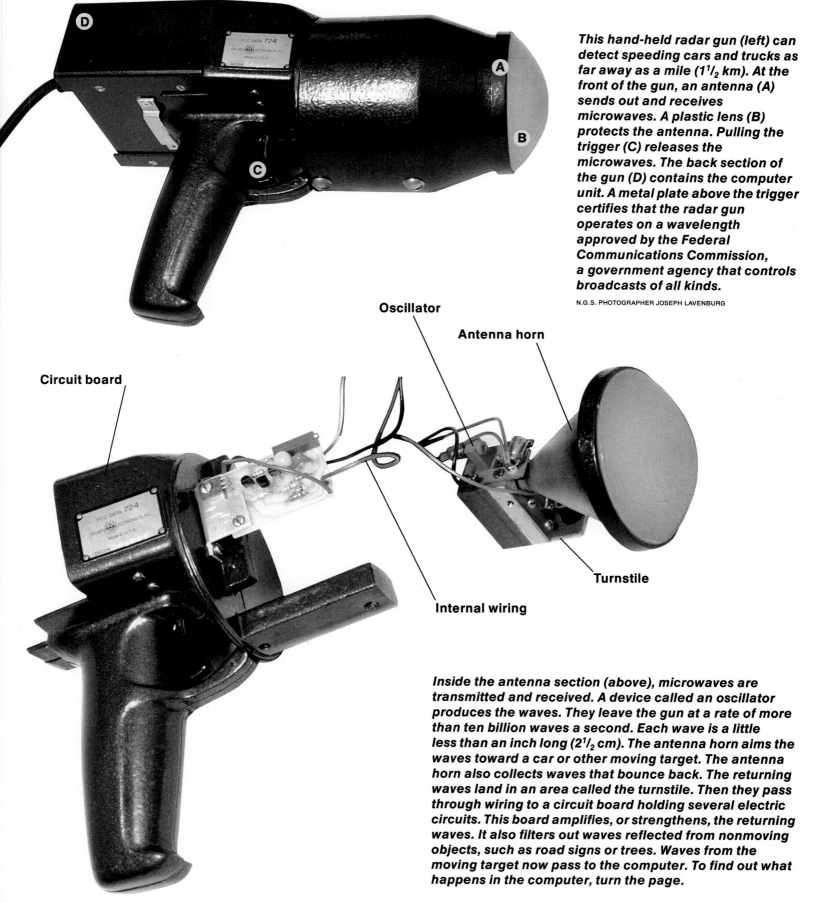

D

A

B

C

This hand-held radar gun (left) can detect speeding cars and trucks as far away as a mile (1¹/₂ km). At the front of the gun, an antenna (A) sends out and receives microwaves. A plastic lens (B) protects the antenna. Pulling the trigger (C) releases the microwaves. The back section of the gun (D) contains the computer unit. A metal plate above the trigger certifies that the radar gun operates on a wavelength approved by the Federal Communications Commission, a government agency that controls broadcasts of all kinds.

N.G.S. PHOTOGRAPHER JOSEPH LAVENBURG

Oscillator

Antenna horn

Circuit board

Turnstile

Internal wiring

Inside the antenna section (above), microwaves are transmitted and received. A device called an oscillator produces the waves. They leave the gun at a rate of more than ten billion waves a second. Each wave is a little less than an inch long (2¹/₂ cm). The antenna horn aims the waves toward a car or other moving target. The antenna horn also collects waves that bounce back. The returning waves land in an area called the turnstile. Then they pass through wiring to a circuit board holding several electric circuits. This board amplifies, or strengthens, the returning waves. It also filters out waves reflected from nonmoving objects, such as road signs or trees. Waves from the moving target now pass to the computer. To find out what happens in the computer, turn the page.

Computer unit

Inside the computer unit (right) a main board (A) gathers the information received by the antenna section. The main board then sends the information through a series of electric circuits located on several other boards. One board (B) determines the frequency of the incoming waves. Another board (C) translates that number into miles or kilometers an hour. Still another (D) stores the speed and displays it on a screen on the computer unit (E). As long as the trigger of the gun is pressed, the speed in the display changes with the speed of the car. When the trigger is released, the displayed speed is locked in. Two buttons tell the computer to check its accuracy. Pressing the lower one (F) should cause "188" to appear in the display. This number means all the parts of the display are working. Pressing the upper one (G) should cause "60" to appear in the display. If it does appear, the computer is making correct measurements.

N.G.S. PHOTOGRAPHER JOSEPH LAVENBURG

Connecting rod to antenna section

Cable to power supply

Radar in sports

At a baseball game, a New York Mets pitcher hurls a ball toward the catcher (left). A radar gun records the speed of the ball in miles an hour (below). Many professional baseball teams now use radar to measure how fast their pitchers throw the ball. This information can help a coach determine when a pitcher is tiring. Team scouts looking for new talent use radar to test the pitching strength of young players. Radar also helps measure speeds in bobsledding and in car, boat, and ski racing.

N.G.S. PHOTOGRAPHER JOSEPH H. BAILEY (ABOVE AND BELOW)

Microwaves from a radar gun bounce back from a moving car (below). The returning waves (gray) differ from the outgoing ones (black). That is because the waves vibrate faster after they bounce off the car. If the car were heading away from the gun, the returning waves would vibrate more slowly. Sound, radio, and light waves also change when they bounce off moving objects. You can hear sound waves change. Ask an adult to drive a car past you while blowing the horn. The horn will have a high pitch as the car nears you. After it passes, the horn will sound lower.

14

How Traffic Signals Work

When you walk down a city street on a quiet evening, you may notice something about the traffic signals. Even when there's no traffic, most signals continue to change: green, yellow, red. A timer inside the signal's control box keeps the signal changing regularly. If you stand beside the box, you may be able to hear the whir of the machinery inside.

Not all traffic signals are preset, however. Some react to sensing devices set in the pavement. Cars activate these devices by rolling over them. Sometimes you may see a signal with a button mounted on the pole. By pushing the button, a person on foot can change the light in order to cross the street.

Controlling traffic flow on busy streets has been a problem in cities and towns for centuries. As long ago as the first century B.C., a Roman leader, Julius Caesar, banned most wheeled vehicles from Rome. In the early 20th century, streets around the world began to fill with cars and trucks. Something had to be done to control their movement and to prevent collisions. Driving laws, road signs, and police officers helped keep the traffic moving smoothly.

In 1914, Cleveland, Ohio, became the first city in the United States to install an automatic traffic signal. By the early 1920s, colored signal lights were blinking throughout the nation. Today's traffic control devices have become complex electronic systems. On these pages, you'll see how some typical signals work.

Streaks of red and white light trace the paths of cars passing through a busy part of Arlington, Virginia, at night. The photographer made this picture by leaving the camera shutter open for a long time while traffic moved past. Signals operate all night at intersections like this one, even when there is little traffic.

The traffic signals pictured at left are regulated by a control box nearby. A timer inside the box operates switches that turn the lights on and off—green for go, yellow for caution, red for stop. The timing cycle has been designed to keep traffic flowing smoothly. It also gives pedestrians—people on foot—enough time to cross the street safely. At some intersections, engineers preset the timer in the control box to change the timing. During rush hour, for example, when many cars crowd the streets, the signal intervals shorten. Late at night, when fewer cars pass, the timer switches from the three-color sequence to a flashing yellow light that tells drivers, "Go ahead, but with caution." The crossing signal at right uses an upraised palm to halt pedestrians. The figure of a man walking tells people it's all right to cross. Visitors unfamiliar with English and children too young to read can understand pictorial signals like these.

A matter of time

During the day, thousands of cars pass through this intersection — the same one shown on page 61. The metal box in the center of the picture contains electronic controls for the traffic lights. In communities with heavy traffic, experts collect information about how many cars use the roads at different times of the day. The number of cars, the speed limit, and the amount of pedestrian traffic help them decide how long each light should stay green, yellow, and red. At this intersection, the yellow signal lasts about four seconds. On streets like these, where the speed

limit is 25 miles an hour (40 km/h), cars and drivers need about four seconds to slow down or to clear the intersection. On a highway where cars travel 55 miles an hour (89 km/h), the yellow-light interval required for a safe stop is about five seconds. Traffic engineers time the lights to allow for the extra braking time.

Rotating timers

A look inside the control box of a traffic signal (above) shows the rotating timers that switch the various colored lights on and off. The devices can be set to operate continuously, day and night, without human help. But the system can also be operated from a distant control center if traffic patterns change. A similar timer will turn a house light on and off automatically when you are away from home. You simply set the timer (right) to a specified time — seven o'clock, for instance. When the rotating dial reaches the hour you have selected, it trips a switch and the light comes on. The lighted windows make any would-be burglars think that someone is at home.

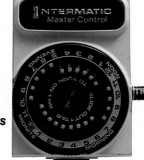

Workers install wire detectors on a side street in Alexandria, Virginia (above). Sealing material will cover the wires, but the wires still will be able to "feel" cars passing over them. When they do, they will notify the traffic signal ahead. In situations like this, where a quiet street meets a busier one, the traffic light changes only when the sensing device signals that a car needs to enter the major street.

Even though traffic signals have become complicated, they still contain simple light bulbs. A traffic department employee (right) changes a burned-out bulb in the yellow light at a busy intersection in Virginia.

Control center: At City Hall in Alexandria, Virginia, traffic specialists watch what's going on outside (above). Their office is the headquarters of a complex computer system that keeps track of street traffic all over the city. Information gathered automatically at every traffic light in Alexandria feeds into this room. At the wall map, a specialist studies the buildup of rush hour traffic. Colored lights tell him which streets have the most cars on them. Another specialist works at a computer terminal. He can use the terminal to change the timing of signals where traffic problems develop during rush hour.

N.G.S. PHOTOGRAPHER JOSEPH H. BAILEY (ALL)

In case of emergency

Sometimes the normal flow of traffic has to be interrupted because of an emergency, such as a fire. If a fire truck or an ambulance must slow down or stop at red lights, the driver loses precious time. Recently, Alexandria, Virginia, installed electronic equipment that its fire department can use to change traffic lights during an emergency. In the picture at right you can see a black and silver box on the roof of a fire truck. It is called a tripping device. Below, look at the circled unit hanging between two traffic lights. It is a special receiver. When a fire truck approaches a red light, the tripping device on the truck sends out a signal that causes the receiver to change the light to green. Then the truck can continue without delay to the scene of the fire. Once the truck passes, the light goes back to its normal preset schedule.

15

How a Calculator Works

What is the sum of 3 + 5? It's 8, of course. How did you come up with the answer? First, your eye saw the numbers and sent signals to your brain. Your brain figured out the question and pulled the answer from your memory. Then it sent signals back that told you to say or write the answer.

A calculator follows similar steps in solving math problems. Its "brain" is a thin piece of silicon (SILL-uh-kun) called a chip. Silicon is a substance that can carry electricity. Thousands of electric switches connected by tiny wires cover the surface of the chip. They make up what is known as an integrated circuit (IC). When you want a calculator to add 3 + 5, you turn on the calculator and press the 3 key. The chip translates the message into electric signals it can understand. The chip then stores the 3 on its surface in a place where new numbers go. Wires link that place to the display area, a rectangle near the top of the calculator. When the "new-number" signals reach the display area, you see a 3.

Next, you press the + key. The chip translates that message into an electric signal and stores it in another place. Now you press the 5 key. The chip moves the 3 from the "new-number" place and stores it somewhere else. The 5 now goes to the "new-number" place, and the number 5 appears on the display. Last, you press the = key. That tells the chip it's time to end the problem. Several steps then take place very rapidly. The chip collects the 3 and the 5. It checks for instructions on what to do with those numbers and finds a +.

It moves the numbers to a spot where they are added together. Programmed instructions in the calculator's memory tell it that 3 + 5 = 8. Eight then goes to the "new-number" place and appears on the display.

The problem is solved!

N.G.S. PHOTOGRAPHER JOSEPH H. BAILEY

Calculators in the classroom speed up problem solving. At Langley Junior High School, in Washington, D. C., eighth-grade math teacher Loretta Bowers guides three students through a calculator lesson (left). Shermontá L. Grant, left, Leonard K. Talbert, and Audrey Adams, all 13, have mastered math basics. Now calculators will take over the routine jobs. This gives the students more time to explore new ways of solving problems.

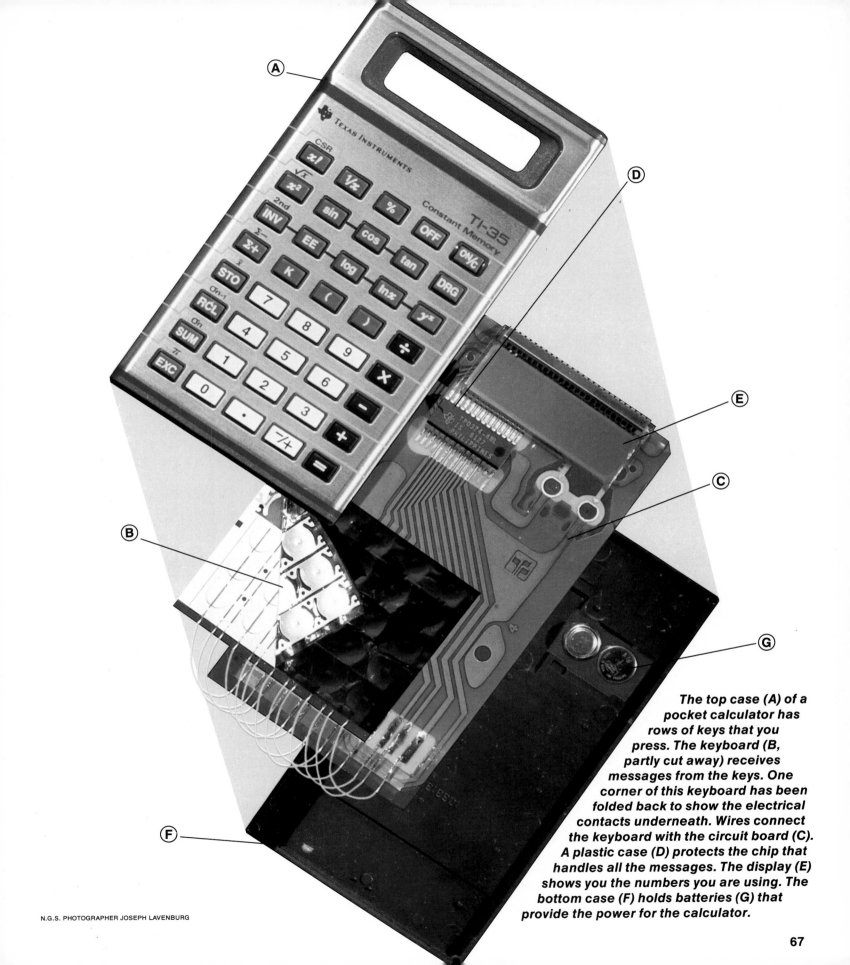

The top case (A) of a pocket calculator has rows of keys that you press. The keyboard (B, partly cut away) receives messages from the keys. One corner of this keyboard has been folded back to show the electrical contacts underneath. Wires connect the keyboard with the circuit board (C). A plastic case (D) protects the chip that handles all the messages. The display (E) shows you the numbers you are using. The bottom case (F) holds batteries (G) that provide the power for the calculator.

N.G.S. PHOTOGRAPHER JOSEPH LAVENBURG

67

On display

Crazy eights! That's all a calculator has in its display area (left). If you have a calculator, move it around until you can see a row of faint eights. Count the segments that make up each eight. There are seven of them. Each segment controls a material called liquid crystal. Wires carry an electric signal from the chip to each segment. When the chip sends a signal to segments of an eight, the liquid crystal between these segments changes. In this drawing, the liquid crystal has darkened. The dark parts show up on the display as numbers. Some calculators may not use liquid crystal, but the control of the segments is the same. This display contains the numbers one through nine. The real calculator display at the left contains only eight numbers. A dot follows each one. When the liquid crystal is changed to darken a dot, the dot becomes a decimal point.

Paths to a chip

A calculator circuit board contains a maze of paths (left) that carry electric signals. All paths lead to the package, a black rectangle that covers and protects a paper-thin chip of silicon. Like a brain, the chip receives, processes, and stores messages from the outside world. It also sends messages back. The surface of the chip contains thousands of microscopic areas that act as switches. These switches are connected to metal pathways. By opening and closing switches, the chip sends electric signals to the right places. They are directed from path to path until they arrive where they belong. The opening and closing of switches produces a pattern of signals somewhat like the dots and dashes of Morse code. This pattern, called a binary code, is the only language a computer understands.

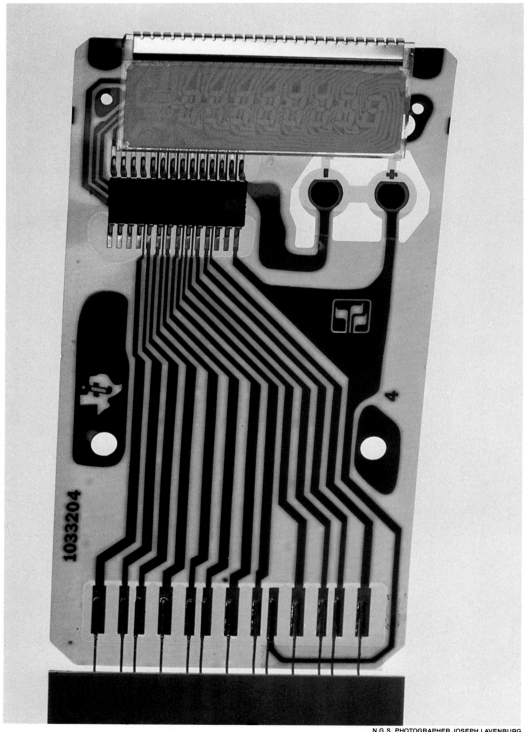

1033204

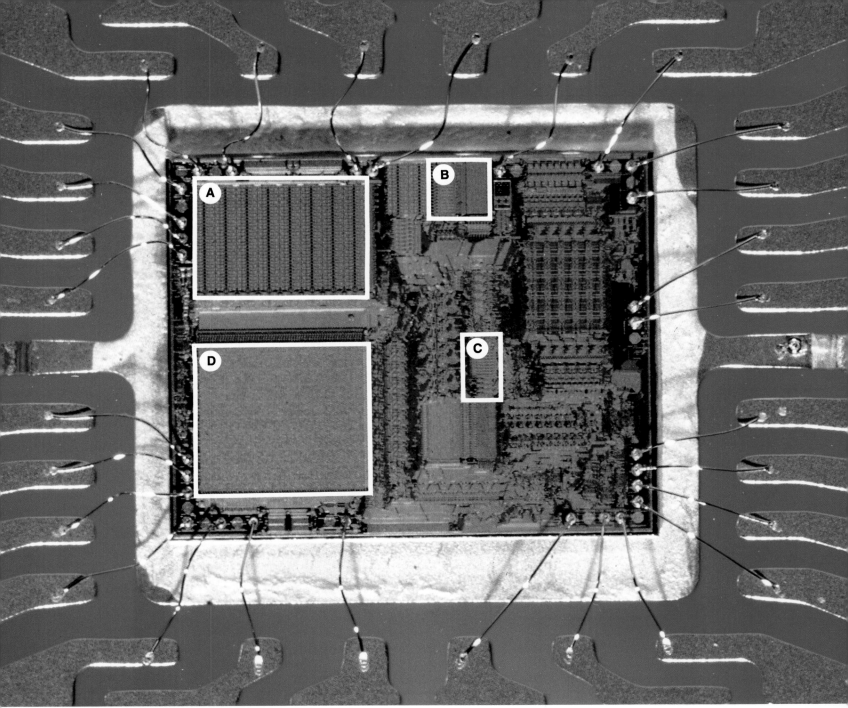

Calculator brain

Magnified about 20 times, a chip shows its crowded surface. Each section has a job to do. A section called the random-access memory, or RAM (A), stores the numbers and operations that you put into the calculator when you press keys. Certain parts store the input numbers while others store the results of the calculations. When you enter a new number into the calculator and it shows up in the RAM, the segment decoder (B) makes sure electric signals get to the right segments, or parts, of the eights in the display area. The adder-subtracter (C) does what its name suggests. It also multiplies and divides. To multiply 15 by 3, it adds 15 three times. To divide 15 by 3, it counts the number of times it can subtract 3 from 15. All of the thousands of instructions that the calculator chip can follow are programmed into a memory called the read-only memory, or ROM (D). The ROM is a permanent memory that cannot be changed. By following these instructions, the calculator can do complicated math problems much faster than a person can. Of course, a person must still figure out how to solve the problem and tell the calculator what to do.

16

How Video Games Work

When PAC-MAN* races through a maze gobbling energy dots and dodging hungry ghosts, he seems almost to come alive. Sound effects and fast-moving displays lend real excitement to video games. The challenging games require quick thinking—and quick reactions.

Computers create the special effects that make today's video games so lively. If you could look inside the console, or case, of a home video game, you'd probably find a special kind of mini-computer called a microprocessor. A microprocessor is a thin chip the size of a baby's thumbnail. It's made of a substance called silicon. Thousands of tiny circuits and switches cover its surface. This tiny chip receives information from three sources. Basic instructions are built into the microprocessor's memory when it is manufactured. They tell it how to receive orders from you and how to create displays on a TV screen. When you put a game cartridge into a slot on the console, the cartridge tells the microprocessor how to play the games in that cartridge. As you play a game, you give the chip additional instructions by moving a joystick or turning a paddle.

All this information must be translated into electric

JAMES A. SUGAR AND PHILIP R. LEONHARDI, N.G.S. STAFF

Tanks tangle on the battlefield of a video game (above). Their commanders are William Allen, 10, left, and Michael Nesuda, 11. Both boys live in Alexandria, Virginia. They are playing a fast-moving game called "Combat." About seven million American homes now have video games. "This game is frustrating, but fun," William says.

Surrounded by computer components, Alan Moss hunts for "bugs" in a video game called "Super Breakout" (left). Moss works as a programmer for Atari Incorporated, in Sunnyvale, California. Programmers create video games. They call a mistake in a game program a "bug." To debug this game, Moss plays it over and over. As he does, he studies the computer printout at his feet. It describes every action in the game. Each time Moss finds a mistake, he enters a correction on the keyboard below the TV screen.*

signals. They are the language a microprocessor understands. As each command signal comes in, the microprocessor searches its memory for the meaning of that signal. To carry out the command, the microprocessor may have to perform complicated calculations. The calculations determine how your move will look on the screen. The electric signals the microprocessor understands then must be converted into a video display—pictures that you understand.

The translating, searching, calculating, and converting take place quickly—in a fraction of a second. The electric signals that do the work travel at nearly the speed of light. They follow thousands of circuits, or pathways, controlled by switches on the chip.

Although home video games are not so complicated as arcade games, all video games work on the same principles and depend on the same modern miracle—the microprocessor. Pictures on these and the following pages give you a look inside home video games.

Dots march across a game printout (above) as a programmer plots the movements of a soccer player for a video game. The programmer uses a special language that he and the computer understand. The dots on the printout stand for electric signals. The signals will trigger the action on the screen. The programmer writes out each move in a game. Then the programmer adds instructions for color and for sound effects, if necessary. Writing the program for a game may take weeks—or even months. After the program is finished and debugged, circuits for it will be designed and put on a chip.

If you took the cover off a game cartridge, this is what you would see (right). The rectangular plastic case in the center shields a memory chip. The chip contains instructions for several versions of one game. Above the chip case, you see a printed circuit board. The circuits carry electric signals from the chip to the microprocessor inside the console. Prongs above the circuit board plug the cartridge into the console.

game program™
PAC-MAN
Use with Joystick Controllers

The brain of a video game lies hidden in the console (above). To play a game, you attach the console to the antenna of a television set. Then you slide a game cartridge into a slot in the console. Depending on the game, players make their moves using joysticks, on the left, or paddles, on the right. A look inside a joystick (right) shows how the stick helps you "talk" with the microprocessor inside the console. The joystick handle tilts in eight directions. It can send eight different electric signals to the microprocessor. Those signals carry your orders about the moves you want to make as you play.

JAMES A. SUGAR (ABOVE AND RIGHT)

A look inside a video-game console reveals a maze of switches, wires, chips, and prongs (below). The game cartridge fits into a slot (A). Six-control switches (B through G) help you set up the game. You turn on the power (B) and tell the console whether you have a black-and-white or color TV set (C). Then you set the desired level of difficulty for the player on the left (D) and for the player on the right (E). You use a switch (F) to select the game in the cartridge that you want to play. You press the reset button (G) if you want to play that game again. Prongs along the top of the console (H) connect it to the joysticks or paddles and to the TV antenna. Section I changes household electric current into current the game can use. Section J converts electric signals produced by the game into TV signals that can be displayed on a screen. The central logic unit (K) processes all the messages from the game cartridge and the players. It sends these messages to the TV screen, where they show up as galloping ghosts or as disappearing dots. Every video-game logic unit contains a microprocessor and one or more additional chips. This video-game console has a three-chip unit.

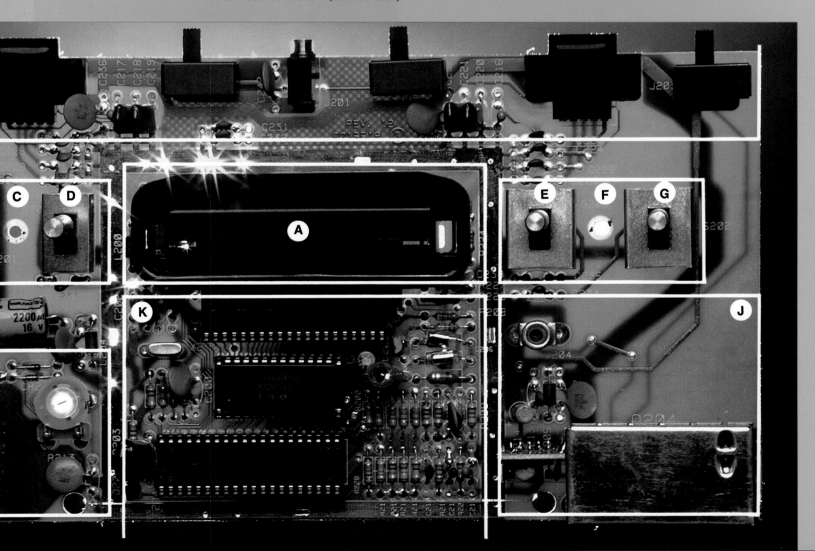

Blueprint of a chip

Moves—and more moves—surround technicians at Atari Incorporated. Daryl Eng, left, and Edwin Chu check the design of a new game cartridge chip. The drawings on the wall and table show electric circuits in the chip. There are *thousands! Each one represents a possible move in a video game. Eng and Chu measure the length of each circuit and check the placement of switches that control the flow of electricity. They make sure that every game move will occur at the right time and in the right place. After they finish, the large drawings will be reduced in size. The pattern will be etched into a chip.*

Sealed in a protective case, a chip rests on a blueprint of its circuit pattern (small inset picture, left). The chip is the tiny square in the center of the rectangular case. Electric wires that provide power for the chip will be connected to small prongs along the edges of the case. The complicated pattern of pathways and switches on this blueprint has been reduced in size to fit on the chip. Several versions of one game may share the same chip. Each game program uses a different pattern of circuits.

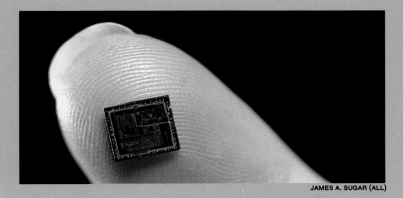

Mighty midget. The chip from a home video-game cartridge rests on a fingertip—with room to spare (above). Chips similar to this one operate electronic watches, pocket calculators, home computers, cash registers, scientific instruments, robots, and many other kinds of electronic devices. They even help engineers design new chips. A chip this size can perform hundreds of thousands of calculations each second. No wonder it's hard to keep ahead of a video game!

Colorful cocoons called resistors, capacitors, and diodes (right) regulate the amount of electric current that flows through a circuit. You saw these tiny regulators on page 73 in section K.

17

How Smoke Detectors Work

You probably have at least one smoke detector in your home—perhaps in a hallway leading to the bedrooms. When you and your family go to sleep at night, you feel secure. You know that if a fire were to break out during the night, the detector should wake you with its loud BE-E-EP alarm. You should have time to escape.

Smoke detectors have become familiar household items in recent years. There are more than 40 million of them in American homes. Some cities have fire regulations that require people to install smoke detectors in their houses. As a result, the detectors have saved many lives and millions of dollars' worth of property.

Have you ever wondered how a smoke detector knows when to let out that piercing warning noise? Smoke detectors operate in several different ways. Here's how one model works:

The smoke detector shown in the photographs on these pages is called a photoelectric detector. This kind of detector contains an area called the smoke-sensing chamber. Inside, a small light source sends a constant beam of light onto a dark surface.

If a fire breaks out, smoke flows into the sensing chamber. The light beam bounces off the smoke and scatters. Some of the light hits a device called a photocell. Photocells are sensitive to light. When light hits the cell, it generates an electric current. The electricity sets off the alarm.

Your smoke detector may have a small light on its cover that flashes on and off at regular intervals. This shows that the detector is operating correctly. It's still a good idea to test the smoke detector once a week. The light in the sensing chamber may need cleaning. Read the instructions that came with your detector. Testing may seem like a bother, but it may help to keep you safe from fire.

Danger! Smoke drifts upward into a smoke detector installed on a ceiling (left). As the smoke enters the detector, it will trigger an alarm. Millions of homes in the United States have smoke detectors in bedrooms, in stairwells, or in hallways. If a fire starts, the loud alarm will warn the people inside the building. It may give them time to escape. Once outside, they can call the fire department.

N.G.S. PHOTOGRAPHER JOSEPH LAVENBURG (ALL)

Smoke-sensing chamber

Alarm

Battery

With its lid removed, a photoelectric smoke detector reveals its inner workings (left). The battery at the bottom of the picture provides power for this detector. The small red globe above the battery is an indicator light. It flashes regularly when the machine is working properly. If the power source for your smoke detector weakens, the alarm—the silver disk in this picture—will go off once every minute. That's the detector's way of saying: "Hey! It's time to change my battery." When you do change it, the alarm stops. The smoke-sensing chamber at the top activates the alarm.

Below, you can see the heart of a smoke detector—the smoke-sensing chamber. The small light source at top right sends out a beam of light. Normally, the beam hits a dark surface on the opposite side of the chamber. But smoke has drifted inside this sensing chamber and the light has scattered. (You can see how this happens by shining a flashlight through the smoke from a campfire.) Some of the light now hits a sensor called a photocell. When this happens, the photocell responds by sending an electric current to the alarm. The alarm sounds its BE-E-EP. If you ever hear that sound, stay calm. Leave the building by a planned fire-escape route, and be grateful that you had a smoke detector to sound the alarm.

Photocell

Light source

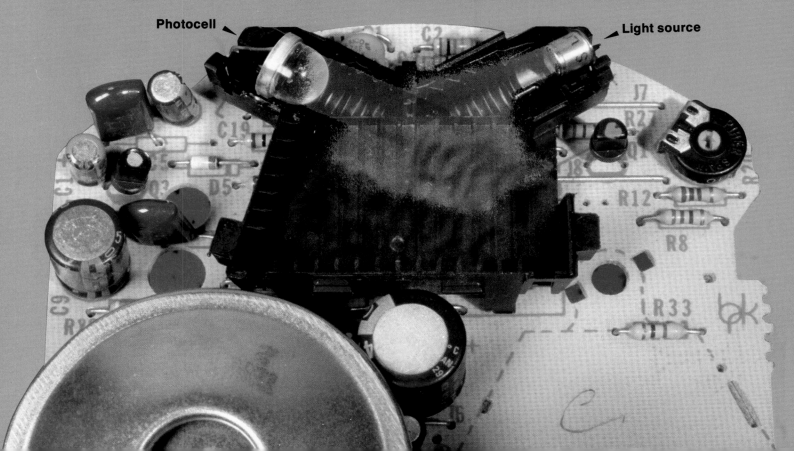

18

How a Fire Sprinkler Works

When you're in a store or a school and notice small metal disks attached to the ceiling, you can feel at ease. These objects are sprinklers. They are there to protect you—and the building—from fire.

If a fire should break out, water or a chemical will spray from the sprinklers directly onto the flames. An alarm will sound, alerting you to the danger. An outside alarm will let the fire department know where to bring equipment to put out the fire.

Sprinkler systems have protected lives and property since the late 1800s. These early systems consisted of pipes with holes punched along their length. The pipes were connected to a water supply. When a fire broke out, someone opened a valve and water ran through the pipes and sprayed out over the fire.

Modern sprinkler systems operate automatically. The fire itself turns on the system. Heat rising from the fire breaks seals, allowing water to spray from pipes.

Today, you can find sprinkler systems in such places as manufacturing plants, hotels, stores, and other buildings that house large numbers of people or large amounts of materials. Company officials often choose this kind of fire protection for this reason: If a fire should break out, only the sprinklers directly above the fire go into action. That means the sprinklers cause no water damage elsewhere in the building.

In some cases, the sprinklers will put out a fire quickly, even before the fire department arrives. In other cases, the sprinklers help contain the fire until fire fighters can put it out. All this means big savings in human lives, in money, and in materials.

Not all sprinkler systems work in the same way. Many have electronic heat sensors that turn them off as well as on. Others are designed to work in locations where water pipes normally would freeze. On these and the following pages, you'll see how the most widely used sprinkler—the wet-pipe system—works.

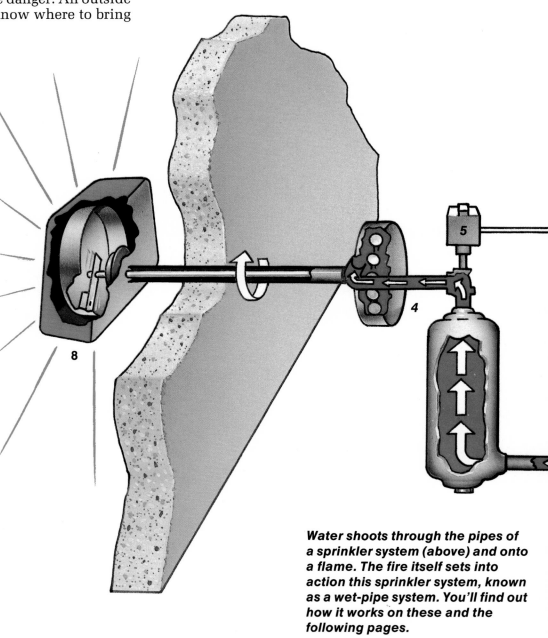

Water shoots through the pipes of a sprinkler system (above) and onto a flame. The fire itself sets into action this sprinkler system, known as a wet-pipe system. You'll find out how it works on these and the following pages.

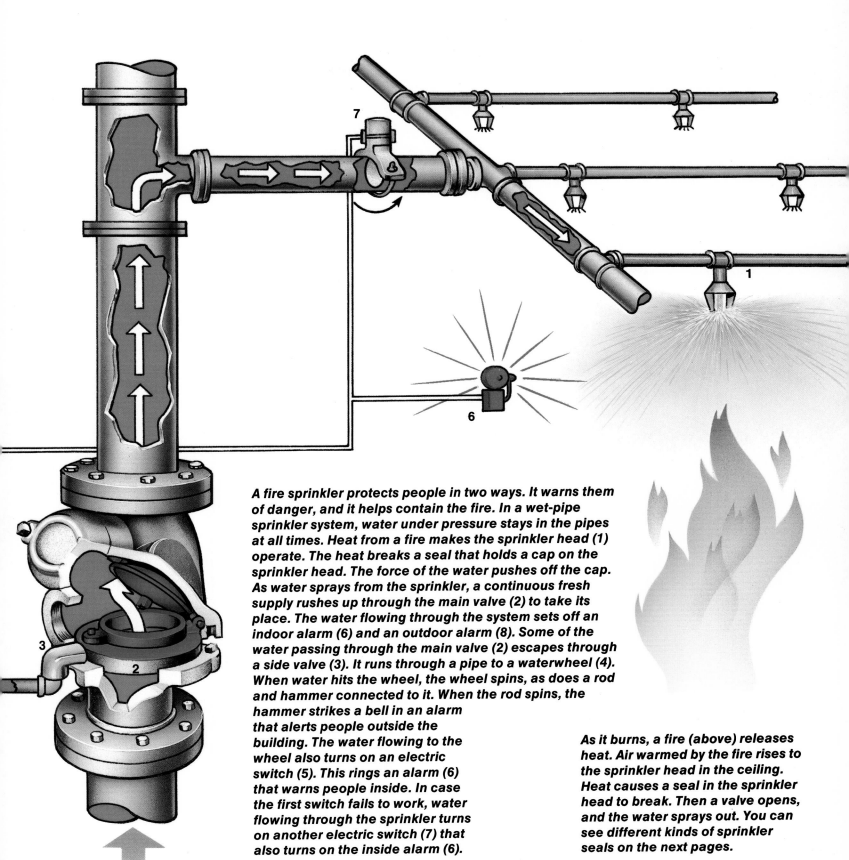

A fire sprinkler protects people in two ways. It warns them of danger, and it helps contain the fire. In a wet-pipe sprinkler system, water under pressure stays in the pipes at all times. Heat from a fire makes the sprinkler head (1) operate. The heat breaks a seal that holds a cap on the sprinkler head. The force of the water pushes off the cap. As water sprays from the sprinkler, a continuous fresh supply rushes up through the main valve (2) to take its place. The water flowing through the system sets off an indoor alarm (6) and an outdoor alarm (8). Some of the water passing through the main valve (2) escapes through a side valve (3). It runs through a pipe to a waterwheel (4). When water hits the wheel, the wheel spins, as does a rod and hammer connected to it. When the rod spins, the hammer strikes a bell in an alarm that alerts people outside the building. The water flowing to the wheel also turns on an electric switch (5). This rings an alarm (6) that warns people inside. In case the first switch fails to work, water flowing through the sprinkler turns on another electric switch (7) that also turns on the inside alarm (6).

As it burns, a fire (above) releases heat. Air warmed by the fire rises to the sprinkler head in the ceiling. Heat causes a seal in the sprinkler head to break. Then a valve opens, and the water sprays out. You can see different kinds of sprinkler seals on the next pages.

MARVIN J. FRYER

Sprinklers operate in several ways. This one (right) contains a glass capsule like those beside it. In this sprinkler, one end of the capsule rests on a water-spreading wheel at the end of the sprinkler. The other end of the capsule presses against a metal cap that seals off the water pipe. Ordinarily, the capsule holds the cap in place; no water can flow from the sprinkler. But when a fire breaks out, heat from the fire causes the capsule to explode, releasing the cap — and the water.

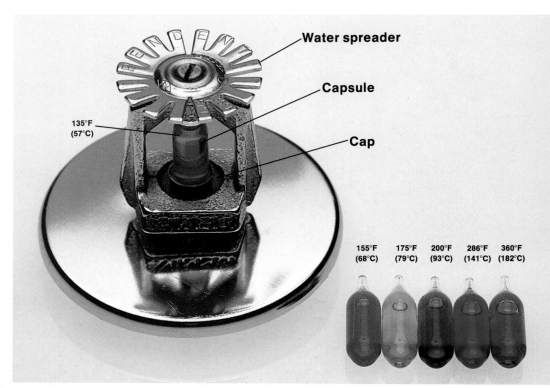

Water spreader

Capsule

Cap

135°F (57°C)

| 155°F (68°C) | 175°F (79°C) | 200°F (93°C) | 286°F (141°C) | 360°F (182°C) |

Water begins to flow as a capsule explodes (above). Heat caused the explosion. Each sprinkler capsule contains a liquid and an air bubble. When heated, the liquid expands, compressing the bubble. When the pressure in the capsule builds beyond a certain point, the capsule shatters. The size of the bubble determines the temperature at which this happens. The larger the bubble, the more heat it takes for the capsule to burst. Colors indicate the temperatures at which each of the capsules at the top of this page will burst.

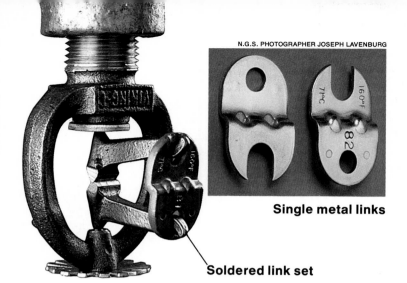

Single metal links

Soldered link set

In another wet-pipe system, two metal links like those above control the sprinkler. The links will be sandwiched together and cemented with a soft metal called solder (SAH-dur). Solder melts at a much lower temperature than does the metal in the rest of the sprinkler. You can see a set of soldered links in place on the sprinkler head beside the single links. The links clip onto the ends of two levers. The levers hold the sprinkler cap in place.

Heat from a torch sets off a link-style sprinkler (below). When the temperature reached 160°F (71°C), the solder cementing the links melted. Then they flew apart and released the ends of the levers. With nothing to hold the levers in place, they flew out. This freed the cap that sealed off the water supply. As the cap falls away, water streams out of the sprinkler head. If you look closely, you can see some of these parts flying from the sprinkler.

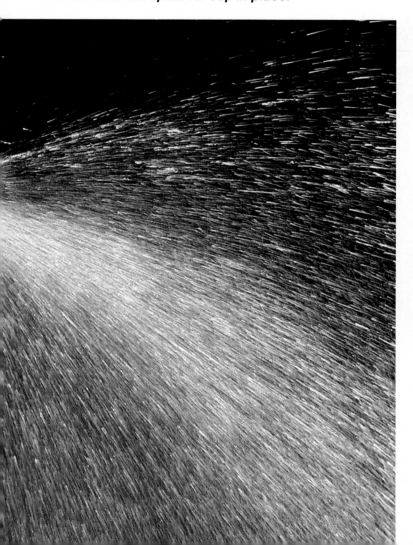

Water flies in all directions as a sprinkler goes off (left). As it leaves the top of the sprinkler, the water pours out in a stream. Then it hits the water spreader. The spreader forces it into an even spray. A sprinkler like this one sprays out 50 gallons of water (189 L) every minute. It will continue to operate until someone turns the water off.

How Sailplanes Work

Imagine how it must feel to soar as birds do, riding the winds high into the sky. In an aircraft called a sailplane, you can do just that. A sailplane is an unusual kind of glider. Like all other gliders, the sailplane has no engine. It glides to the ground on air currents. Unlike other gliders, however, the sailplane can climb high into the sky on updrafts—rising air currents.

Columns or bubbles of heated air form some of these updrafts. They are called thermals. Another kind of updraft is known as a ridge lift. A ridge lift forms when winds sweep up and over a mountain ridge. If the pilot can find many strong updrafts, a sailplane can travel hundreds of miles before it finally glides to a landing.

Since a sailplane has no engine, it first must be towed into the sky. Once released, gravity—the natural pull of the earth against the weight of the plane—begins to pull it to the ground. As the plane moves, air passing over the top of the curved wings pushes down on them. The pressure of air traveling under the wings is greater, so this air pushes up. This upward force is called lift. By working against the pull of gravity, lift helps the plane glide gradually to the ground. The forward movement that results is called thrust.

But as the sailplane glides, the moving air also pushes against its surfaces, creating a force called drag. Drag slows the plane. Drag also pushes against the movable surfaces of the wings and tail, helping the pilot to steer.

The streamlined shape of a sailplane provides it with a great amount of lift and with little drag. That enables it to glide for great distances before it touches the ground. A draft of rising air adds even more lift to the sailplane. When the lift is greater than the weight of the plane, the sailplane can soar.

DOUG HEWINS (BOTH)

Pilot trainees tackle ground lessons at the Black Forest Gliderport, near Colorado Springs, Colorado (above). They are, from the left, François Webner, 14, of Van Wert, Ohio; Brad Hindman, 14, of Castle Rock, Colorado; Stephen Whitaker, 17, of Wilton, Connecticut; and Keith Nickerson, 15, of Platte City, Missouri. Ground lessons cover everything from weather analysis to flight techniques. "There's a lot to learn," says Stephen. "You can't just hop in the plane and go." At 14, a trainee can apply for a license to learn how to fly a sailplane.

Charles Cook, 15, of Fort Worth, Texas, watches for clouds that might signal thermals, updrafts of heated air (right). His instructor, Tony Fernandez, coaches him from behind. "Up in the air I feel just like a bird," says Charles. "The only sound is the wind whistling by." This craft is a two-seater training model. The backseat has controls just like those in the front seat. Fernandez can easily correct any errors Charles might make.

Set for takeoff, Keith closes the canopy, or cockpit hatch (left). As required by law, he wears both a shoulder harness and a safety belt. A sailplane has no engine, so a tow plane will pull Keith and Mark Palmer, his instructor, into the sky. A tow rope links the sailplane to the tow plane ahead. The tow plane will take Keith to an altitude of about 2,000 feet (610 m). At the same time, the tow pilot will look for an updraft that Keith's sailplane can ride. When an instrument in Keith's cockpit shows he is in an updraft, Keith will release the tow rope.

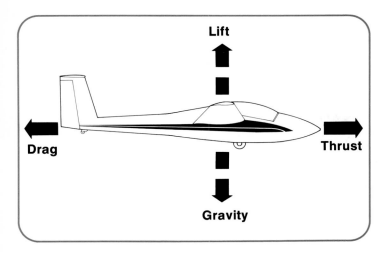

Lift

Drag

Thrust

Gravity

A sailplane moves through the air because gravity pulls on the weight of the plane (left). But as the plane falls, air moving past the wings produces a force that offsets some of the pull of gravity. This offsetting force is called lift. When gravity and lift combine, the plane moves forward. This forward movement is called thrust. But as the sailplane moves forward, air pushes against it, creating a force called drag. Drag reduces the thrust of the plane.

Brad receives a traditional soaking after his first solo flight (below). At Black Forest Gliderport, trainees usually solo — fly alone — within ten days. Most fly at least 30 times with an instructor before they solo.

Soaring on sun power

Thermals help sailplanes stay in the air (left). The sun causes the thermals. The sun heats the surface of the earth unevenly. Areas such as plowed fields, rooftops, beaches, and parking lots absorb heat more quickly than other areas do. As these spots heat up, they heat the air directly above them. Warm air is lighter than cool air, so the heated air rises. High in the sky, it cools. This causes it to release moisture it has carried up with it. The moisture forms puffy clouds. Thermals are, of course, invisible. So pilots look for the clouds that cap them. Then they glide from one thermal to another, left. As they circle inside each thermal, they are carried up with the rising air. Sailplanes also can ride along a ridge lift, far left. A ridge lift forms when wind shoots up and over a rising slope.

The sailplane's design helps it
use the air more efficiently than any
other flying machine does. Its long,
slender wings create much lift and little
drag. Its narrow fuselage (FEW-suh-lahj), or
body, reduces drag. A comparison of lift and drag
tells a pilot how far a plane can glide. This comparison is called the glide ratio. A sailplane's glide ratio is
very high. It may reach 60 to 1. That means that if the
sailplane were towed to an altitude of 1 mile ($1\frac{1}{2}$ km),
the sailplane would glide 60 miles (97 km), even without using updrafts, before reaching the ground. Of
course, a pilot must be able to control the plane in flight.
A sailplane has special instruments and movable parts
to help the pilot do this. One of the most important instruments is the variometer (vah-ree-AHM-uh-tur). The
variometer measures changes in air pressure—or air
weight—as the sailplane gains or loses altitude. Light
air signals an updraft; heavy air signals a downdraft.

*Instruments fill a sailplane's cockpit (below). They tell the
pilot the sailplane's altitude, its airspeed, the speed at
which it is climbing or descending, and the direction it is
going. The handle at left controls the dive brakes. The red
knob releases the tow rope. Floor pedals and an upright
control stick enable the pilot to steer.*

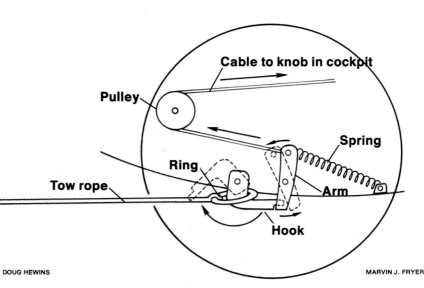

Cable to knob in cockpit

Pulley

Spring

Ring

Arm

Tow rope

Hook

DOUG HEWINS

MARVIN J. FRYER

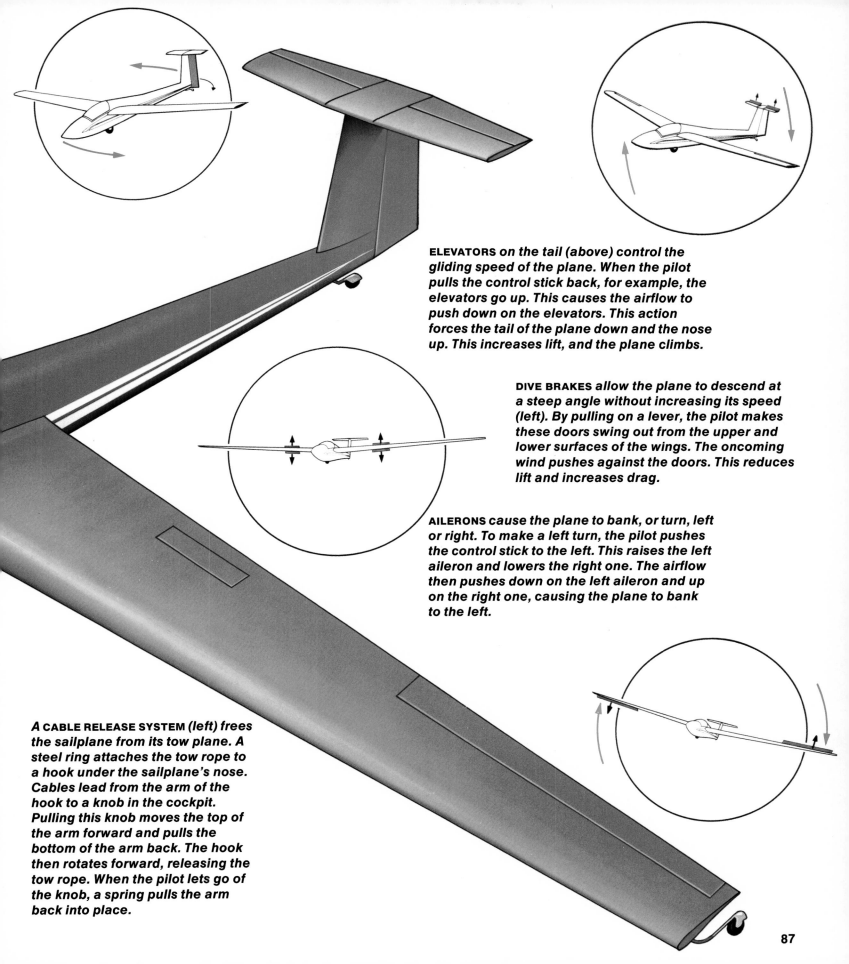

ELEVATORS *on the tail (above) control the gliding speed of the plane. When the pilot pulls the control stick back, for example, the elevators go up. This causes the airflow to push down on the elevators. This action forces the tail of the plane down and the nose up. This increases lift, and the plane climbs.*

DIVE BRAKES *allow the plane to descend at a steep angle without increasing its speed (left). By pulling on a lever, the pilot makes these doors swing out from the upper and lower surfaces of the wings. The oncoming wind pushes against the doors. This reduces lift and increases drag.*

AILERONS *cause the plane to bank, or turn, left or right. To make a left turn, the pilot pushes the control stick to the left. This raises the left aileron and lowers the right one. The airflow then pushes down on the left aileron and up on the right one, causing the plane to bank to the left.*

*A **CABLE RELEASE SYSTEM** (left) frees the sailplane from its tow plane. A steel ring attaches the tow rope to a hook under the sailplane's nose. Cables lead from the arm of the hook to a knob in the cockpit. Pulling this knob moves the top of the arm forward and pulls the bottom of the arm back. The hook then rotates forward, releasing the tow rope. When the pilot lets go of the knob, a spring pulls the arm back into place.*

20

How Hot-air Balloons Work

They drift with the breeze through the sky—huge, colorful aircraft called hot-air balloons. Each is made of only a handful of parts: nylon fabric; cables; a gondola, or basket; and a burner or two fueled by propane gas.

Such a craft rises because the hot air inside its envelope, or bag, is lighter than the cooler air outside it. The heavier air keeps the balloon afloat much the way water keeps a boat afloat. The balloon will rise as long as the air inside it is hotter, and therefore lighter, than the outside air.

With its gondola in place beneath the envelope, a balloon can carry people. But a great deal of lift is needed to handle the additional weight. To produce enough lift, the air inside the envelope must reach a certain temperature. How high that temperature is depends on the weight of the load, the temperature of the outside air, and the size of the envelope. On an average day with an average load, the envelope air temperature would need to be at least 100°F higher than the outside temperature. Otherwise the balloon would never get off the ground. An instrument called a pyrometer (pie-ROM-uh-ter) keeps track of the air temperature at the top of the envelope. To hold the balloon in level flight or to make it rise higher, the pilot must reheat the air inside the envelope from time to time by turning on the burners for a few seconds.

The pilot also controls the balloon's descent. The envelope has vents that open or close when lines are pulled from inside the gondola. The vents allow some of the heated air to escape. Then the air inside the envelope cools and the balloon starts downward. What a hot-air balloon pilot can't do is control the craft's speed or its movement forward, backward, or from side to side. Wind currents determine these. The pilot keeps moving the balloon up or down and searches for the best currents to ride.

Preparing to launch a hot-air balloon, members of a ground crew heat the air inside its nylon bag, or envelope, with a propane torch (left). Minutes before, they blew the air in with a motorized fan. As the air heats, the envelope will slowly rise to an upright position. Ropes hold the balloon down until pilot and passengers board.

BOHDAN HRYNEWYCH/PICTURE GROUP

Balloons in flight offer a bird's-eye view of the countryside near Albuquerque, New Mexico (right). Before a launch, most balloonists release a small helium balloon to check wind direction and speed. Winds stronger than 8 miles an hour (13 km/h) make launching and landing too dangerous. At first, the crew members drift high, to check the wind currents. Then they usually descend to 300 or 400 feet (91 or 122 m) to enjoy the scenery. With practice, a balloonist can estimate wind speed by studying the motion of flags and smoke rising from smokestacks (inset).

N.G.S. PHOTOGRAPHER OTIS IMBODEN (RIGHT)   ADAPTED BY MARVIN J. FRYER, COURTESY OF DR. WILL HAYES, AUTHOR OF *THE COMPLETE BALLOONING BOOK* (INSET)

0-1 mph (0-1½ km/h)   1-3 mph (1½-5 km/h)   4-7 mph (6-11 km/h)   8-12 mph (13-19 km/h)

Coming down

Every hot-air balloon has an opening at the top called the deflation port. One type of port (left) is known as a parachute top because of its design. Ropes (green lines) attach the edges of the parachute top to a control line (red). By pulling on this line, the pilot can control the downward movement of the balloon. The pilot opens the port a little to make minor adjustments in altitude. Used this way, the port does the same job as the side vent shown in the drawing on the next page. If the pilot opens the port more, the balloon descends faster. To close the port, the pilot releases the cord. Air inside the envelope quickly pushes the panel shut.

ADAPTED BY MARVIN J. FRYER. COURTESY OF DR. WILL HAYES, AUTHOR OF *THE COMPLETE BALLOONING BOOK*

Another type of deflation port is the rip top (left and opposite). Self-sticking tape attaches the circular panel to the top of the envelope. The rip line (the dotted red line) runs from one edge of the port to the side of the envelope and down to the gondola. When the pilot pulls this cord, the vent peels open from the side. The rest of the time, the tape — as well as air pressure within the envelope — keeps the port shut. To make a fast landing, the pilot "rips out" — opens the port entirely. Then air escapes from the envelope within minutes. Balloons with rip tops have side vents as well. The side vents help the pilot make small changes in altitude.

MARVIN J. FRYER (LEFT AND OPPOSITE)

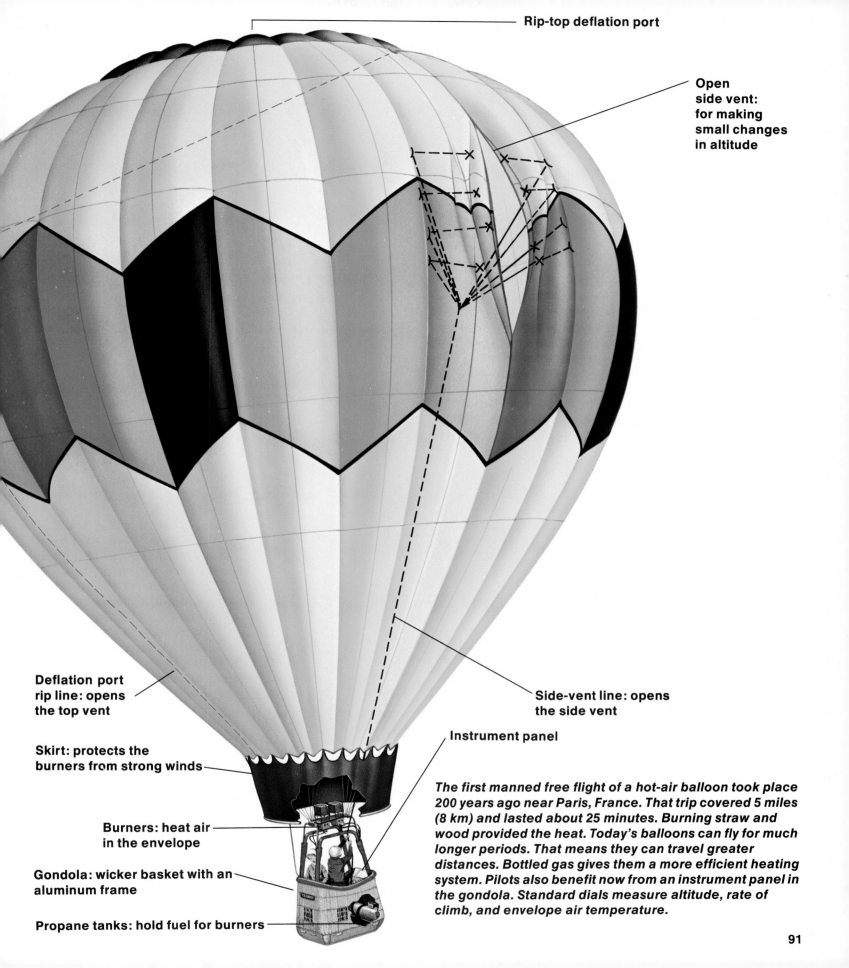

Rip-top deflation port

Open side vent: for making small changes in altitude

Deflation port rip line: opens the top vent

Side-vent line: opens the side vent

Instrument panel

Skirt: protects the burners from strong winds

Burners: heat air in the envelope

Gondola: wicker basket with an aluminum frame

Propane tanks: hold fuel for burners

The first manned free flight of a hot-air balloon took place 200 years ago near Paris, France. That trip covered 5 miles (8 km) and lasted about 25 minutes. Burning straw and wood provided the heat. Today's balloons can fly for much longer periods. That means they can travel greater distances. Bottled gas gives them a more efficient heating system. Pilots also benefit now from an instrument panel in the gondola. Standard dials measure altitude, rate of climb, and envelope air temperature.

21

How the Space Shuttle Works

Look! Up in the sky. It's a bird. It's a plane. No, it's the space shuttle—the world's first vehicle to take off like a rocket, orbit the earth like a spacecraft, then land like an airplane. It shuttles between earth and space. It is designed to be used over and over, just as trains, buses, and planes are used on trip after trip.

When the shuttle blasts off, it has four parts. A cargo plane called the orbiter carries a huge bullet-shaped fuel tank on its belly. Attached to either side of the tank are two booster rockets. The booster rockets and the orbiter's own engines lift the bulky shuttle away from the pull of the earth's gravity. About eight minutes after lift-off, the orbiter will fly alone. The empty boosters and the fuel tank will have fallen back to earth.

The orbiter needs only a small push from two on-board rockets to put it into orbit. When it returns from space, the orbiter has no power. It depends on its airplane parts to position it as it glides in for a landing.

Spacecraft rockets behave very much like the rockets people shoot off on the Fourth of July. A firework rocket flies into the air when the explosive powder inside it burns, producing a gas. The rocket cylinder doesn't have enough room inside to hold the gas, so the gas shoots out a hole at the bottom. As it shoots out, the gas pushes against the cylinder and lifts the rocket up. Rocket engines work the same way, but on a much larger scale. By controlling the direction of its rocket fire, a spacecraft can steer and even brake itself.

Computers control the firing of the shuttle's rockets. Five computers act as the shuttle's "brain." Through a variety of programs, they give the craft commands during launch, flight, and landing. The astronauts and the ground team at Mission Control can also give the shuttle commands, but the computers actually carry out the commands.

Four computers take charge of the mission. The fifth serves as a backup, in case something goes wrong with one of the other four. All process the same information and each one can fly the orbiter. If something does go wrong, an alarm sounds and a light on the control panel lights up to identify the problem. Then the astronauts use keyboards to ask the computers to describe the problem so a solution can be worked out. These computers are similar to those often used in schools and homes, but the memory of each one on the shuttle is many times greater.

JON T. SCHNEEBERGER, N.G.S. STAFF

Piggy-backing into space, the orbiter Columbia *lifts off (right). It is attached to its huge fuel tank. The orbiter is the part of the space shuttle that looks like an airplane. Two long booster rockets on either side of the orbiter provide the push to get the craft off the launch pad and beyond the pull of the earth's gravity. The boosters fire for only two minutes. During that short burst they produce enough thrust, or forward force, to lift 25 fully loaded 747 airplanes into the air. When empty, the boosters parachute back to earth and are recovered later. About six minutes into the flight, the main engines on the orbiter shut down. The big fuel tank falls away, never to be used again. The orbiter* Enterprise *(left) was not designed to fly in space. It was used in testing procedures in preparation for the* Columbia *flight.*

A payload area 60 feet (18 m) long provides space to store scientific equipment inside the orbiter. In this photograph (right), taken in orbit, the crew has opened the payload bay doors. A mission's payload is its cargo of operating machinery. This area can carry more than 30 tons (27 t) of cargo, including satellites, space telescopes, and smaller items. For a minimum of $3,000, space can be rented in the payload area for an approved experiment.

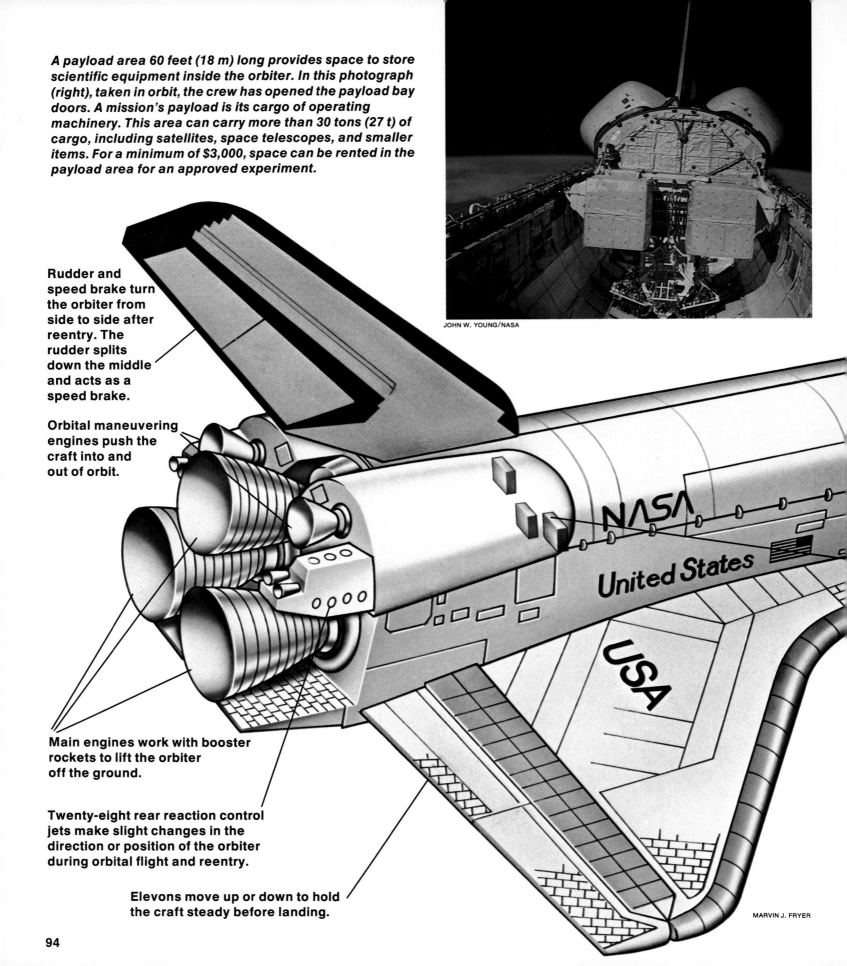

JOHN W. YOUNG/NASA

Rudder and speed brake turn the orbiter from side to side after reentry. The rudder splits down the middle and acts as a speed brake.

Orbital maneuvering engines push the craft into and out of orbit.

Main engines work with booster rockets to lift the orbiter off the ground.

Twenty-eight rear reaction control jets make slight changes in the direction or position of the orbiter during orbital flight and reentry.

Elevons move up or down to hold the craft steady before landing.

MARVIN J. FRYER

94

Joe Engle joins fellow astronaut Richard Truly, far right, in the cockpit of an orbiter model. If you think the flight panel looks complicated, you're right. Dials indicate such things as airspeed and altitude. Keyboards send orders to computers. Three TV screens display computer data. Astronauts train for an average of 1,300 hours on models like these before they fly the shuttle. These astronauts used their training when they flew the Columbia on its mission in November 1981.

Sixteen forward jet engines mounted in the nose form the forward part of the orbiter's reaction control system. Together with the rear engines, they help change the direction of the ship in space. The forward jets do not operate during reentry.

These six compartments contain the avionics (a-vee-AH-nix) system. The system includes navigation instruments and a group of five computers that do most of the actual flying of the shuttle. The computers are located in the forward compartments. They control the distribution of electric power to all of the orbiter's systems. They process all the information that comes from instruments aboard the craft, compare the results, and relay commands to the various parts of the orbiter. The avionics system also controls the communications systems and keeps track of the orbiter's position at all times during the flight.

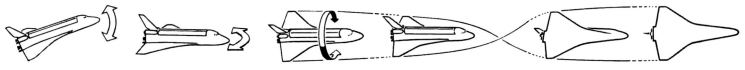

Pitch	Yaw	Roll		Roll Reversals

Moving in space

Computers control the 44 reaction control jet engines on the orbiter. These engines change the direction of the orbiter in space. Computers fire the engines in different combinations to push the craft in different directions. They can make the craft pitch—swing its nose up or down—or yaw—twist to the left or to the right. They can also put it into a roll, or a roll reversal, as the drawings above show. Roll reversals, or S-curves, slow the orbiter down before it comes in for a landing.

To space and back

From fiery start to high-speed finish, the space shuttle makes a smooth flight. As you follow the drawings on these pages from left to right, you'll see how the orbiter behaves as it makes its journey. After lift-off, the computers release the boosters. Then the computers roll the craft over on its back to release its empty fuel tank. Then they command the orbiter's own orbital maneuvering system (OMS) engines to push the craft into orbit. There, the orbiter maintains a stable course for the mission. When the time comes to return home, the reaction control jets turn the orbiter tailfirst. Firing in this direction, the OMS engines act as a brake, pushing the craft out of orbit. The orbiter drops into the earth's atmosphere and glides to a landing.

▲ The shuttle maintains a stable course in its orbits around the earth. On this imaginary mission, it moves from west to east at 25 times the speed of sound. It will remain in orbit for 8 days. Small pushes from the reaction control jets keep the craft on course — or make adjustments.

◄ The orbiter drifts upside down for about a minute. Then the orbital maneuvering system engines fire and push the craft into orbit.

◄ Just before the orbiter goes into orbit, it rolls on its back and releases the now empty external fuel tank. The tank falls away, breaking up as it goes. The pieces land harmlessly in a remote part of the ocean.

Two minutes after launch, the boosters run out of fuel and separate from the tank. They parachute to earth, to be used again. The main engines continue to push the shuttle toward the spot where it will begin to orbit the earth. They operate for a total of about eight minutes of flight. Their fuel comes from a large tank attached to the orbiter's belly.

Ignition! All three main engines in the orbiter fire within seconds of each other. Two rocket boosters, attached to the external tank, fire almost immediately after the main engines fire. The combined thrust of the engines pushes the shuttle off the ground.

▼

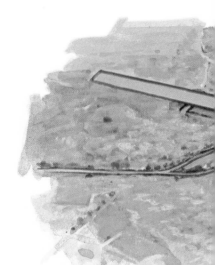

MARVIN J. FRYER

The cargo bay doors open so the crew can complete experiments. Opening the doors also allows excess heat from the electrical system to escape. The rear and forward reaction control jets fire to turn the plane around. It will leave orbit tailfirst.

The orbital maneuvering engines fire. With the craft moving tail first, the engine firing slows the craft enough to let it fall out of orbit.

Forward and rear reaction control jets maneuver the orbiter into a nose-high position to prepare it for reentry into the earth's atmosphere. The jets will position the orbiter so that protective tiles on the belly of the craft will absorb heat caused by friction of the air hitting the orbiter when it enters the atmosphere.

Thirty minutes before touchdown, the orbiter enters the atmosphere. Ten minutes later, it has reached the period of maximum heating. Its belly is red-hot. During this time, the crew cannot communicate with ground control; electrically charged particles surrounding the craft interfere with radio transmissions.

The orbiter begins to slow down for its landing by going through a series of S-curves. Ten minutes before touchdown, a combination of three guidance systems goes into operation. This system is designed to land the orbiter automatically. However, the astronauts can land the craft themselves, if they choose.

NORTH AMERICA

Edwards Air Force Base

Kennedy Space Center

PACIFIC OCEAN

EQUATOR

SOUTH AMERICA

ROBERT W. NORTHROP

From its take-off at the Kennedy Space Center in Florida, the shuttle speeds into space. It enters orbit — a flight path about 173 miles (278 km) above the earth, where the speed of the orbiter offsets the pull of gravity. The drawing above shows the path of one complete orbit, and the landing site at Edwards Air Force Base, in California.

Because of its size and weight, the orbiter glides at a steep angle. The craft is completely powerless now. It depends on its elevons, rudder, and speed brake to adjust position and speed. Thirty seconds before touchdown, the commander pulls up the nose to level the orbiter for the final approach to the landing strip. Within seconds the space shuttle will complete another successful mission.

97

With its landing gear down and a chase plane keeping watch, the Columbia II orbiter comes in for a landing (left). During the last minutes of reentry, the orbiter crew takes over from the flight computers. The commander uses the elevons and the rear reaction control jets to steer the craft through a series of rolls. He uses these rolls to slow the orbiter down, much the way a skier makes wide turns to slow down on a steep slope. During the last 30 seconds before landing, the commander pulls the orbiter's nose up and locks the landing gear into place. Below, the Columbia III orbiter touches down after a mission.

MICHAEL STEINBERG (INSET)
GLENN CRUICKSHANK/WEST STOCK, INC.

Glossary

acoustics—a branch of physics concerned with the study of sound

amplifier—a device that strengthens electric current or power

amplitude modulation (AM)—the changing of the height from top to bottom of a radio carrier wave

antenna—a device, usually a metal rod or wire, that sends or receives electromagnetic waves

aperture—the opening in the diaphragm of a camera lens that allows light to pass through

atoms—small particles consisting of a nucleus and one or more electrons that make up all substances

audio-frequency signals (AF)—the electric signals that a microphone creates from sound waves

calculator—an electronic counting device that uses an integrated circuit to do all of its figuring

carrier waves—high frequency radio waves that radio stations combine with audio-frequency signals for broadcasting

chip—see integrated circuit

derailleur—the part of a multispeed bicycle that shifts gears by moving a chain from one rear sprocket to another or from one chainwheel to another

diaphragm—in a camera, a series of overlapping metal plates that control the amount of light entering through the lens; in a speaker, a thin, flexible disk that vibrates in response to incoming electric signals and helps produce sounds

electromagnet—a wire or coil of wire that behaves like a magnet while an electric current flows through it

fluorescent—having the ability to give off light when energy is absorbed from another source

frequency—the number of complete vibrations a sound wave or an electromagnetic wave makes in a second

frequency modulation (FM)—the moving closer or spreading apart of the vibrations of a radio carrier wave

fuselage—the central body of an airplane or spacecraft that contains the cockpit and that supports the wings and the tail

geographic poles—the two points on the earth that mark the ends of the earth's imaginary axis

gravity—the force that pulls against the weight of an object

integrated circuit—a small piece of silicon containing thousands of microscopic electric switches and connecting wires that control the flow of electric signals

lens—a device that collects and focuses light

magnetic field—the invisible force that exists between the two poles of a magnet

magnetic poles—the two areas of a magnet where the magnetic force or the magnetic field is the strongest

mainspring—the spring that is the power source for some mechanical watches and clocks

mechanics—the branch of physics concerned with the motion of objects and the forces that act upon them

microprocessor—an integrated circuit that provides in one chip the functions equal to those carried out by the central processing unit of a computer

optics—the branch of physics concerned with light and the effects of forces acting upon it

photocell—a device that converts light into an electric signal

photons—tiny particles of energy that make up all light waves

physics—the science that studies matter and energy and how they interact

random-access memory (RAM)—the section of an integrated circuit, or chip, that temporarily stores and retrieves information while the chip is being used

read-only memory (ROM)—the section of an integrated circuit, or chip, that contains information permanently stored during the manufacture of the chip

silicon—a natural substance found in quartz rock that can carry electric signals

silicon chip—see integrated circuit

solar wind—a force of electrically charged particles that flow from the sun

solder—a soft metal used to join metallic surfaces

sprocket—a wheel ringed with teeth onto which fit such things as film or a chain

thermals—rising air currents caused by columns or bubbles of heated air

thermodynamics—the branch of physics concerned with heat

updrafts—rising air currents

Additional Reading

Readers may want to check the National Geographic Index and the WORLD Index in a school or a public library for related articles and to refer to the following books: ("A" indicates a book for readers at the adult level.)

General Books
Graf, Rudolf and George Whalen, *How It Works, Illustrated: Everyday Devices and Mechanisms*, Van Nostrand Reinhold Company, 1974 (A). Lazarus, David and Manfred Raether, *Practical Physics: How Things Work*, Stipes Publishing Co., 1979 (A). Kerrod, Robin, *The Way It Works*, Mayflower Press, 1980. Lambert, Lye, Taylor and Wicks, *All Color Book of Science Facts*, Arco Publishing, Inc., 1980. Pollard, Michael, *How Things Work*, Larousse and Co., Inc., 1978. *The Illustrated Science and Invention Encyclopedia*, H. S. Stuttman, Inc., 1982 (A).

Books on Specific Subjects
Cuthbertson, Tom, *Anybody's Bike Book*, Ten Speed Press, 1979. Sloane, Eugene A., *The All New Complete Book of Bicycling*, Simon & Shuster, Inc., 1981. Paula Z. Hogan, *Inventions that Changed Our Lives: the Compass*, Walker & Company, 1982. Weiss, Harvey, *Motors and Engines and How They Work*, Harper & Row Publishers, Inc., 1969. Ayliffe, Jerry, *American Premium Guide to Coin Operated Machines*, Crown Publishers, Inc., 1981. Bostrom, Roald, *Cameras*, Raintree Publishers, Inc., 1981. Stern, Rudi, *Let There Be Neon*, Harry N. Abrams, Inc., 1979. Johnson, Jim, *A Look Inside Lasers*, Raintree Publishers, 1981. Kettelkamp, Larry. *Lasers, the Miracle Light*, William Morrow & Company, 1979. Benade, Arthur H., *Horns, Strings, and Harmony*, Greenwood, 1979 (A). Walther, Tom, *Make Mine Music!*, Little, Brown and Company, 1981. Kettelkamp, Larry, *The Magic of Sound*, William Morrow and Company, 1982. Lewis, John, *Pocket Calculator*, EDC Publishing, 1981. Olney, Ross and Pat, *Pocket Calculator Fun and Games*, Franklin Watts, Inc., 1977. Cohen, Daniel, *Video Games*, Archway Paperbacks, 1982. D'Ignazio, Fred, *Electronic Games*, Franklin Watts, Inc., 1982. Smith, Norman F., *Gliding, Soaring, and Skysailing*, Julian Messner, 1980. Adler, Irene, *Ballooning High and Wild*, Troll Associates, 1976. Wirth, Dick and Jerry Young, *Ballooning: The Complete Guide to Riding the Winds*, Random House, Inc., 1980. Fichter, George S., *The Space Shuttle*, Franklin Watts, Inc., 1981. Priestley, Lee, *America's Space Shuttle*, Julian Messner, 1978.

Index

CONSULTANTS

Joseph H. Hamilton, Ph.D., Vanderbilt University; W. Edward Lear, Ph.D., American Society for Engineering Education — *Chief Consultants*
Glenn O. Blough, LL.D., University of Maryland; Suzanne F. Clewell, Ph.D., University of Maryland — *Educational Consultants*

Nicholas J. Long, Ph.D., University of Michigan — *Consulting Psychologist*

The Special Publications and School Services Division is grateful to the individuals, organizations, and agencies named or quoted within the text and the individuals cited here for their generous assistance:

Hank Albright, Fawn Engineering Corporation
Nazir A. Ali, Nurion, Inc.
Atari Incorporated
David Bauer, BRK Electronics
Richard E. Berg, University of Maryland
L. H. Bradley, Viking Corporation
Lois Burke, Decatur Electronics, Inc.
William B. Clare, Baldwin Piano & Organ Company
Lee Dickinson, National Transportation Safety Board
Leong Dong
Lorraine Doran, New York Mets
James Douglass, Coin Acceptors, Inc.
Bohdan P. Duma
Bernard S. Finn, Smithsonian Institution
William R. Fonda, George Washington University
Sidney Forrest, Peabody Conservatory of Music
Phyllis Gallahan
Lou Gorman, New York Mets
James R. Harman, General Electric Company

Ray Hawkins, General Electric Company
Mel Heider, Westclox
Wallace D. Henderson, BDM Corporation
S. Henriksen, National Geodetic Survey
Jack Huff, Baldwin Piano & Organ Company
Peter Huggler, Toastmaster, Inc.
R. F. Kerzaya, Fairfax County Police Department
Henry Klein, Black and Decker Manufacturing Company
Haynes A. Lee, Laser Institute of America
Gerald Luecke, Texas Instruments Incorporated
Steven P. McKeown, Schweizer Aircraft Corporation
Paul Mock, Selmer Company
Thomas L. Moser, NASA Johnson Space Center
Chan Oakman, Insurance Company of North America
George A. Painter, Black Forest Gliderport
Panasonic Company
Anthony Peritore, National Geographic Society Staff
John R. Reynolds, Co-op Vendors, Inc.
Roger Sawtelle, Rockville Bicycle Shop
Paul C. Schultheiss, Stanley Automatic Openers
F. L. Shupe, Fairfax County Police Department
John Simpkinson, Baldwin Piano & Organ Company
Peter B. Stifel, University of Maryland
Terry White, NASA Johnson Space Center
David Wiegers, Decatur Electronics, Inc.
Bill Wildprett, Silva Compass
William C. Wilkinson III, Bicycle Manufacturers Association of America, Inc.
F. A. Wittern, Sr., Fawn Engineering Corporation

Library of Congress CIP Data
Main entry under title:

How things work.

(Books for world explorers)
Bibliography: p.
Includes index.
   SUMMARY: Uses such familiar objects as a bicycle, neon sign, calculator, and hot air balloon to introduce the field of physics.
   1. Physics — Juvenile literature.   2. Technology — Juvenile literature.   [1. Physics. 2. Technology]   I. National Geographic Society (U. S.)   II. Series.
QC25.H65   1983   600   81-47894
ISBN 0-87044-425-5 (regular binding)
ISBN 0-87044-430-1 (library binding)

HOW THINGS WORK
PUBLISHED BY
THE NATIONAL GEOGRAPHIC SOCIETY
WASHINGTON, D. C.

Gilbert M. Grosvenor, *President*
Melvin M. Payne, *Chairman of the Board*
Owen R. Anderson, *Executive Vice President*
Robert L. Breeden, *Vice President, Publications and Educational Media*

PREPARED BY THE SPECIAL PUBLICATIONS AND SCHOOL SERVICES DIVISION

Donald J. Crump, *Director*
Philip B. Silcott, *Associate Director*
William L. Allen, William R. Gray, *Assistant Directors*

STAFF FOR BOOKS FOR WORLD EXPLORERS

Ralph Gray, *Editor*
Pat Robbins, *Managing Editor*
Ursula Perrin Vosseler, *Art Director*

STAFF FOR HOW THINGS WORK

Margaret McKelway, *Managing Editor*
Suzanne Nave Patrick, *Researcher and Project Editor*
Charles E. Herron, *Picture Editor*
Drayton Hawkins, *Designer*
Marvin J. Fryer, *Artist*
Ross Bankson, Roger B. Hirschland, Merrill Windsor, *Contributing Editors*
Sharon L. Barry, William P. Beaman, Jan Leslie Cook, James A. Cox, Sandra Lee Crow, Jacqueline Geschickter, Bruce Lewenstein, Michael Lipske, Catherine O'Neill, Mortimer P. Reed, *Writers*
Donna B. Kerfoot, *Researcher*
Jane R. Halpin, *Editorial Assistant*
Artemis S. Lampathakis, *Illustrations Assistant*
Janet A. Dustin, *Art Secretary*

STAFF FOR FAR-OUT FUN!

Patricia N. Holland, *Project Editor;* Jane R. McGoldrick, *Text Editor;* Ursula Perrin Vosseler, *Designer;* Jan Watkins, *Artist*

ENGRAVING, PRINTING, AND PRODUCT MANUFACTURE

Robert W. Messer, *Manager;* George V. White, *Production Manager;* Richard A. McClure, *Production Project Manager;* Mark R. Dunlevy, David V. Showers, Gregory Storer, *Assistant Production Managers;* Katherine H. Donohue, *Senior Production Assistant;* Mary A. Bennett, *Production Assistant;* Julia F. Warner, *Production Staff Assistant*

STAFF ASSISTANTS: Nancy F. Berry, Rebecca Bittle, Pamela A. Black, Nettie Burke, Mary Elizabeth Davis, Rosamund Garner, Victoria D. Garrett, Nancy J. Harvey, Joan Hurst, Katherine R. Leitch, Virginia W. McCoy, Mary Evelyn McKinney, Cleo Petroff, Victoria I. Piscopo, Tammy Presley, Sheryl A. Prohovich, Carol A. Rocheleau, Kathleen T. Shea, Katheryn M. Slocum

MARKET RESEARCH: Mark W. Brown, Joseph S. Fowler, Carrla L. Holmes, Meg McElligott Kieffer, Susan D. Snell

INDEX: Teresa S. Purvis

Composition for HOW THINGS WORK by National Geographic's Photographic Services, Carl M. Shrader, Director; Lawrence F. Ludwig, Assistant Director. Printed and bound by Holladay-Tyler Printing Corp., Rockville, Md. Color separations by NEC, Inc., Nashville, Tenn. *Classroom Activities Folder* produced by Mazer Corp., Dayton, Ohio.